北京历史文化遗产保护传承体系规划
（2023年—2035年）

Master Plan for Conservation and Inheritance of Beijing's Historic and
Cultural Heritage (2023-2035)

BEIJING

北京历史文化名城保护委员会　编
Beijing Historic City Conservation Commission

中国建筑工业出版社
CHINA ARCHITECTURE & BUILDING PRESS

审图号：GS京（2025）0808号

图书在版编目（CIP）数据

北京历史文化遗产保护传承体系规划：2023年–2035
年 = Master Plan for Conservation and Inheritance
of Beijing's Historic and Cultural Heritage（
2023–2035）：英汉对照 / 北京历史文化名城保护委员
会编. -- 北京：中国建筑工业出版社，2025.5.
ISBN 978-7-112-31049-4

Ⅰ. TU984.21

中国国家版本馆CIP数据核字第2025TF4371号

责任编辑：徐　冉　刘　丹
书籍设计：锋尚设计
责任校对：王　烨

北京历史文化遗产保护传承体系规划（2023年—2035年）

Master Plan for Conservation and Inheritance of Beijing's Historic and Cultural Heritage (2023-2035)

北京历史文化名城保护委员会　编
Beijing Historic City Conservation Commission

*

中国建筑工业出版社出版、发行（北京海淀三里河路9号）
各地新华书店、建筑书店经销
北京锋尚制版有限公司制版
北京中科印刷有限公司印刷

*

开本：787毫米×1092毫米　1/16　印张：9　插页：2　字数：114千字
2025年5月第一版　　2025年5月第一次印刷
定价：**49.00**元
ISBN 978-7-112-31049-4
（44727）

北京历史文化遗产保护传承体系规划
（2023年—2035年）

工作统筹：

首都规划建设委员会办公室

北京历史文化名城保护委员会办公室

技术支持：

北京市城市规划设计研究院

中国建筑设计研究院有限公司建筑历史研究所

Master Plan for Conservation and Inheritance of Beijing's Historic and Cultural Heritage
（2023-2035）

Coordinated by:

Office of Capital Planning and Construction Commission

Office of Beijing Historic City Conservation Commission

Technical support by:

Beijing Municipal Institute of City Planning and Design

Institute of Architectural History, China Architecture Design & Research Group Co., Ltd.

序

党的十八大以来，以习近平同志为核心的党中央把历史文化保护传承工作摆在更加突出的位置。习近平总书记十分关心北京历史文化遗产保护工作，指出"北京是世界著名古都，丰富的历史文化遗产是一张金名片，传承保护好这份宝贵的历史文化遗产是首都的职责。""北京历史悠久，文脉绵长，是中华文明连续性、创新性、统一性、包容性、和平性的有力见证。"习近平总书记的重要论述，为做好北京历史文化遗产保护传承工作提供了根本遵循、指明了方向。

党的二十大报告提出："加大文物和文化遗产保护力度，加强城乡建设中历史文化保护传承……坚守中华文化立场，提炼展示中华文明的精神标识和文化精髓，加快构建中国话语和中国叙事体系。"

2021年8月，中共中央办公厅、国务院办公厅印发《关于在城乡建设中加强历史文化保护传承的意见》，明确"建立城乡历史文化保护传承体系三级管理体制。国家、省（自治区、直辖市）分别编制全国城乡历史文化保护传承体系规划纲要及省级规划"。2022年10月，中共中央、国务院印发《全国国土空间规划纲要（2021—2035年）》，要求"保护传承文化与自然价值，彰显国土空间魅力"。为深化落实上述重要文件要求，结合《北京城市总体规划（2016年—2035年）》实施工作，北京历史文化名城保护委员会组织编制《北京历史文化遗产保护传承体系规划（2023年—2035年）》（以下简称《保护传承体系规划》）。

《保护传承体系规划》坚持以历史文化价值为导向，全面保护利用古代与近现代、城市与乡村、物质与非物质等历史文化遗产，突出展示中华文明历史、中国近现代历史、中国共产党历史、中华人民共和国成立与发展历程、改革开放和社会主义现代化建设的伟大征程，系统保护传承各类历史文化遗产，延续历史文脉，推动城乡建设高质量发展，助力全国文化中心建设再上新台阶。

《保护传承体系规划》作为专项规划，是指导北京历史文化遗产整体保护和可持续发展的工作指南，是制定实施历史文化遗产保护传承利用相关规划方案、政策措施和工程项目的重要依据。

PREFACE

Since the 18th National Congress of the Communist Party of China (CPC), the CPC Central Committee with Comrade Xi Jinping at its core has placed greater emphasis on the preservation and inheritance of history and culture. General Secretary Xi Jinping has paid great attention to the conservation of Beijing's historic and cultural heritage, stating that "Beijing is a world-renowned ancient capital, and its rich historic and cultural heritage is a shining calling card. It is the responsibility of the capital to protect and pass on this precious cultural heritage." He further emphasized, "Beijing has a time-honored history and a continuous cultural lineage, serving as a powerful testament to the continuity, innovation, unity, inclusiveness and peaceful nature of Chinese civilization." General Secretary Xi Jinping's important remarks provide fundamental guidance and a clear direction for the work of protecting and passing on Beijing's historic and cultural heritage.

The report to the 20th National Congress of the Communist Party of China states: "Increase efforts in the conservation of cultural property and cultural heritage, strengthen the preservation and inheritance of history and culture in urban and rural construction... Stay firmly rooted in Chinese culture, collect and refine the defining symbols and best elements of Chinese culture, and accelerate the development of a Chinese discourse and narrative framework."

In August 2021, the General Office of the CPC Central Committee and the General Office of the State Council issued the *Guidelines on Strengthening the Preservation and Inheritance of Historical and Cultural Heritage in Urban and Rural Construction*, which clearly outlined "the establishment of a three-tier management system for the preservation and inheritance of history and culture in urban and rural areas. It specified that "the State should develop a national planning outline for the preservation and inheritance of history and culture in urban and rural areas and provincial master plans should be respectively formulated at the provincial level (as well as autonomous regions and municipalities directly under the central government)". In October 2022, the CPC Central Committee and the State Council issued the *National Territorial Spatial Planning Outline (2021-2035)*, which called for "the preservation and transmission of cultural and natural values, highlighting the charm of territorial land and space." To further implement the requirements of these important documents and in line with the implementation of the *Beijing Urban Master Plan (2016-2035)*, the Beijing Historic City Conservation Commission organized the compilation of the *Master Plan for Conservation and Inheritance of Beijing's Historic and Cultural Heritage (2023-2035)* (hereinafter referred to as the "Master Plan").

The Master Plan follows a historical and cultural value-oriented approach, aiming to comprehensively protect and valorize historic and cultural heritage from ancient to modern times, in both urban and rural areas, and at both tangible and intangible dimensions. It emphasizes exhibiting the history of Chinese civilization, the history of modern China, the history of the Communist Party of China, the founding and development of the People's Republic of China, and the heroic journey of reform and opening-up, and socialist modernization. The Master Plan is a systematic document for protecting and passing on various types of historic and cultural heritage, preserving historical and cultural lineages, promoting high-quality development in urban and rural construction, and contributing to the further advancement in building Beijing into a national cultural center.

The Master Plan, as a sector planning, serves as a guideline for the overall conservation and sustainable development of Beijing's historic and cultural heritage. It provides an important basis for preparing and implementing related planning schemes, policy initiatives, and development projects for the conservation, inheritance, and utilization of historic and cultural heritage.

目 录

| 第四章 |

构建传承体系，弘扬中华文化

| 第五章 |

完善实施机制，保障有效管理

附图

CONTENTS

| **Chapter 4** |

Establishing an Inheritance Mechanism to Promote Chinese Culture

| **Chapter 5** |

Improving the Implementation Mechanism to Ensure Effective Management

Annexed Maps

第一章

————

总则

一、指导思想

坚持以习近平新时代中国特色社会主义思想为指导，全面贯彻党的二十大和二十届二中、三中全会精神，深入学习贯彻习近平文化思想，落实习近平总书记对北京重要讲话和指示批示精神，坚定文化自信，坚持守正创新，立足有利于突出中华文明历史文化价值，有利于体现中华民族精神追求，有利于向世人展示全面真实的古代中国和现代中国，保护第一、传承优先，构建北京城乡历史文化遗产保护传承体系，全面真实讲好中国故事、中国共产党故事、北京故事，支撑中华文明标识体系建设。牢固树立新发展理念，正确处理文化遗产保护与利用、文化遗产保护与经济社会发展的关系，统筹保护传承利用，推动中华优秀传统文化创造性转化和创新性发展，厚植中国式现代化的文化底蕴，为赓续中华文脉、建设社会主义文化强国贡献北京力量。

二、主要规划依据

1.《高举中国特色社会主义伟大旗帜为全面建设社会主义现代化国家而团结奋斗——习近平在中国共产党第二十次全国代表大会上的报告》（2022年）

2.《中共中央办公厅 国务院办公厅关于在城乡建设中加强历史文化保护传承的意见》（2021年）

3.《全国国土空间规划纲要（2021—2035年）》（2022年）

4.《中共中央办公厅 国务院办公厅关于加强文物保护利用改革的若干意见》（2018年）

5.《中华人民共和国文物保护法》（2024年修订）

6.《中华人民共和国非物质文化遗产法》（2011年）

7.《历史文化名城名镇名村保护条例》（2017年修订）

8.《北京历史文化名城保护条例》（2021年）

9.《北京城市总体规划（2016年—2035年）》（2017年）

10.《首都功能核心区控制性详细规划（街区层面）（2018年—2035年）》（2020年）

三、时空界定

本次规划范围与北京市行政辖区范围一致，总面积为16410平方公里。

本次规划期限为2023年至2035年，其中近期至2027年，远期至2035年。

四、规划原则

坚持统筹谋划，系统推进。坚持党委领导、政府统筹，加强顶层设计，建立分类科学、保护有力、管理有效的北京城乡历史文化遗产保护传承体系，做到时间全贯通、空间全覆盖、要素全囊括、全流程保护、全社会参与。统筹规划、建设、管理，促进历史文化遗产保护传承与城乡建设融合发展。

坚持价值导向，应保尽保。用遗产的眼光看、从文明的角度论，凝练北京历史文化遗产核心价值，支撑中华文明标识体系建设，为加快构建中国话语和中国叙事体系提供有效支撑。按照真实性、完整性的保护要求，适应历史文化遗产保护特征，在城乡建设中树立和突出中华文化符号和中华民族形象，弘扬和发展中华优秀传统文化、革命文化、社会主义先进文化。

坚持合理利用，传承发展。坚持以人民为中心，将保护传承融入经济社会发展、生态文明建设和现代生活，充分发挥历史文化遗产的文化价值、社会价值和宣传教育作用。鼓励积极合理使用，促进创造性转化、创新性发展，注重民生改善，不断满足人民日益增长的美好生活需要。

坚持多方参与，形成合力。鼓励和引导社会力量广泛参与保护传承工作，充分发挥市场作用，激发人民群众参与的主动性、积极性，形成有利于历史文化遗产保护传承的体制机制和社会环境。

五、规划目标

到2027年，梳理北京城乡历史文化遗产，持续丰富保护名录，构建北京城乡历史文化遗产保护传承体系，形成一批可复制、可推广的保护利用经验，历史文化遗产保护传承工作融入城乡建设，助力首都高质量发展。

到2030年，完成各类历史文化资源普查和保护名录认定，初步建成北京城乡历史文化遗产保护传承体系。历史文化遗产保护传承工作全面融入城乡建设。北京在中华文明标识体系中的保护重点更加突出，北京历史文化遗产核心价值的阐释和展示得到加强，大国之都的文化影响力、传播力、亲和力进一步提升。

到2035年，全面建成北京城乡历史文化遗产保护传承体系。历史文化遗产保护传承工作深度融入城乡建设和经济社会发展大局。建成彰显文化自信与多元包容魅力的世界文化名城，突出展现中华文明的连续性、创新性、统一性、包容性、和平性。

六、主要任务

一是凝练北京城乡历史文化遗产的价值特征，支撑中华文明标识体系建设，构建基于价值的历史文化遗产保护传承整体格局。

二是明确城乡历史文化遗产的保护传承要点，坚持保护第一、传承优先，以价值阐释为核心，构建完整保护系统，创新传承路径。

三是加强保护管理，完善保护实施机制。衔接国家重大战略及国家经济社会发展规划、国土空间规划等相关规划，明确保护传承推荐项目。

第二章

———┼———

凝练价值体系，引领保护传承

一、人文历史脉络梳理

北京地处北纬40°线附近，位于中国地形第二级阶梯和第三级阶梯的交界地带、400毫米等降水量线和农牧交错地带东南侧。北京小平原地处中国华北平原北端，是内蒙古高原、华北平原和东北地区三大地理单元的交汇之地。

北京位于温带季风性大陆气候区，拥有复杂多样的地形地貌和气候条件，西倚太行山，北靠燕山，永定河、潮白河、北运河、泃河和拒马河五大水系孕育了丰富的生态资源。

北京保存了旧石器时代最早的人类活动遗迹，周口店遗址及周边地区发现了距今约70万年前的北京猿人、10万～20万年前的新洞人、4万年前的田园洞人、1万～3万年前的山顶洞人遗迹，涵盖了直立人、早期智人、晚期智人三个古人类阶段。新石器时代早期的门头沟东胡林人遗址（距今9000～11000年）是粟作起源遗址之一。新石器时代中期及晚期的平谷上宅遗址（距今6000～7500年）出土的黍、粟淀粉粒及工具实证了该时期北京地区已出现旱作农业。

北京地处多种生业区与多民族交汇之地，历经两周时期华夏文明圈边缘的诸侯国都城、秦汉以来的边疆重镇、辽金以来的国家都城。北京琉璃河遗址是考古发现的北京地区最早的城址，一系列出土材料证明了《史记》中对西周初年"封召公于北燕"的记载，"太保墉匽"铭文明确了西周燕都的筑城者为太保召公奭。自此以后，城市史3000余年未曾中断。至1153年金朝迁都燕京并改称中都，北京首次成为统治范围囊括中原核心地区的北方民族王朝的都城，开启了北京870余年建都史。至元朝，北京作为统一王朝的国都，跃升为全国的政治中心、文化中心，这一定位在明清时期持续得到加强。

北京是洋务运动、戊戌变法的策源地，新文化运动和五四运动的全国中心，马克思主义在中国早期传播主阵地，中国共产党主要孕育地之一，在中国现代化进程中具有不可替代的重要历史地位。

北京在新中国成立后完成了从传统社会都城向现代共和国首都的转型，在党的领导下开展以"十大建筑"、天安门广场改造等为代表的首都建设；改革开放后，北京启动了城市经济转型；20世纪90年代，北京进一步强化了全国政治中心和文化中心的城市性质；2008年北京夏季奥林匹克运动会显著提升了中国的国际影响力；党的十八大以来，党中央对北京提出"四个中心"的城市战略定位，即全国政治中心、文化中心、国际交往中心、科技创新中心。以北京城市副中心建设、2022年北京冬季奥林匹克运动会举办为标志，北京逐步迈向中华民族伟大复兴的大国首都、国际一流的和谐宜居之都。

二、多视角的价值体系框架构建

中华文明源远流长。规划着眼于百万年人类史、1万年文化史、5000多年文明史、180多年近现代史、100多年建党史、70多年新中国史、40多年改革开放史，在全国统一的价值体系框架指导下，从政治、经济、社会、科技文化、地理生态等多元视角系统梳理北京在中华文明历史发展脉络中的重要地位，构建北京历史文化遗产核心价值体系，梳理北京城乡历史文化保护重点，重构基于价值的北京城乡历史文化遗产保护体系；通过与国际、国内同类古都型城市对比研究，凸显北京独特的历史文化特征。

三、北京历史文化遗产核心价值体系

（一）整体价值综述

北京历史悠久，文脉绵长，是中华文明连续性、创新性、统一性、包容性、和平性的有力见证。

北京独特的地理格局与地质特征，是东亚地区早期人地关系的杰出见证；北京是中国古代城市营建的集大成者，是见证中华文明赓续的千年古都；北京是近现代中国政体转制的伟大见证，是引领中国式现代化的首善之都；北京见证了欧亚大陆不同文明、不同民族、不同文化之间的碰撞、交流与融合，是多元交流、开放包容的大国之都。

与国际同时期国都相比，北京是礼制式都城规划思想的典范，代表东方文明的最高成就。

（二）价值主题及特征

1．价值一：东亚地区早期人地关系的杰出见证

北京地区位于中国北方第二级阶梯与第三级阶梯的交界地带，保存有丰富的地质景观、人类起源与农业起源遗迹、野生动植物资源，是见证东亚地区早期人地关系的杰出范例。

价值特征Ⅰ–1：北京以周口店遗址为代表的"北京人"遗迹在人类演化史上具有里程碑意义。

价值特征Ⅰ–2：北京以东胡林人遗址为代表的遗址遗迹可为东北亚地区的粟作农业起源提供杰出的见证。

价值特征Ⅰ–3：北京房山、延庆等区域的地质遗迹是中国北方喀斯特地貌地质构造和燕山运动的最典型代表，西部和北部山区丰富的野生动植物资源展现了华北地区的生物多样性。

2. 价值二：见证中华文明赓续的千年古都

北京是公元12世纪以降中华文明传统社会后期金、元、明、清的国家都城所在，是中国古代都城规划理想与礼制空间秩序演变的最终形态，代表了这一时期东亚地区城市文明的最高成就。

北京所依托的燕山山脉介于大兴安岭与太行山山脉之间，拥有中国地理文化上的重要地位，交通区位可沟通各方，连接农耕、游牧、渔猎等多种生业地带，形成"南控中原、北连朔漠"的战略态势，承载着统一的多民族国家都城人地关系的地理基础；在不同时期的空间向背影响下形成的长城防御体系与城市布局，见证了不同民族和文化的冲突与交融。

价值特征Ⅱ-1：北京的地理区位优势在中国古代国家大一统过程中具有独特的战略控扼价值，是统一的多民族国家定都北京的重要地理基础。

价值特征Ⅱ-2：北京自西周以来的城市营建，及中国古代社会后期金、元、明、清不同朝代对北京的都城建设，是中华文明统一的多民族国家发展的伟大见证。

价值特征Ⅱ-3：北京的老城空间规划格局体现了以礼仪制度为核心的中国古代都城规制经典范式。

价值特征Ⅱ-4：北京的古代礼制建筑群格局展现了中华文明在构建与维系传统社会秩序方面的系统性和代表性。

价值特征Ⅱ-5：北京的古代宫殿建筑、皇家寺观、皇家园林与陵寝是中国古代皇家建筑的最高典范，是完整展现中国传统宫廷文化的系列杰作。

价值特征Ⅱ-6：北京的长城体系是中国历史上农耕文化与游牧

文化、渔猎文化碰撞与融合过程的独特见证。

价值特征Ⅱ-7：北京的运河体系见证了中国古代统一的多民族国家漕运制度发展和一系列水利工程技术创新，保障了北京作为统一的多民族国家都城的资源供给。

3．价值三：引领中国式现代化的首善之都

北京近现代历经了中国数千年来未有之变局，是马克思主义在中国早期传播的主阵地和中国共产党的主要孕育地之一；20世纪初以来，中国从传统走向现代的社会转型过程中见证了近现代政治体制转变、思想文化转变、社会性质与经济模式转变、城市职能与发展模式转变，由封建帝国都城走向共和国首都，成为引领中国式现代化进程的首善之都。

价值特征Ⅲ-1：北京近现代经历了政权更迭、思想碰撞、革命历程、城市变化，见证了从传统社会都城向现代共和国首都的转型。

价值特征Ⅲ-2：新中国成立后，北京在中国共产党领导下开展的首都建设，成为中国人民推进社会主义革命和建设的突出示范。

价值特征Ⅲ-3：改革开放后，首都发展迈出新步伐、"四个中心"功能持续增强，凸显了北京在以中国式现代化推进中华民族伟大复兴进程中的引领作用。

4．价值四：多元交流、开放包容的大国之都

北京千年来历经契丹、女真、蒙、汉、满等不同民族政权的建设经营，各民族及各区域的风俗文化、宗教信仰、科技思想充分碰撞、交流与融合，最终形成了兼收并蓄的城市多元文化特征，展现了开放包容的大国国都精神。

价值特征Ⅳ-1：北京是中国多民族文化交流融合的荟萃地，展

现了大国之都的开放包容。

价值特征Ⅳ-2：北京的佛教、道教、伊斯兰教、基督教、天主教等不同宗教信仰和民间传统信仰遗存展现了大国之都的兼收并蓄。

价值特征Ⅳ-3：北京的胡同、合院、会馆、商肆、茶馆、戏楼、村落、农业和手工业等场所及其历史环境、历史植被等展现了独特的京味文化传统。

价值特征Ⅳ-4：北京随着东西方思想文化、科学技术等交流而出现的各类公共建筑与城市设施，见证了社会生产生活方式与城市发展模式的转变。

价值特征Ⅳ-5：北京现存的元代以来中国历代最高学府和教育文化机构遗存及诸多名人旧（故）居，见证了中国教育制度的变迁，是多元文化思想的集大成者。

第三章

强化系统保护，筑牢资源本底

一、加强保护名录管理

（一）完善保护对象体系

北京城乡历史文化遗产保护传承体系的保护对象以具有保护价值、承载不同历史时期文化内涵的城市和村镇、自然和人文、物质和非物质等遗产及其赋存环境为主体，主要包括：（1）世界文化遗产；（2）文物；（3）历史建筑［包括优秀近现代建筑、工业遗产、挂牌保护院落、名人旧（故）居等］和革命史迹；（4）历史文化街区、特色地区和地下文物埋藏区；（5）历史文化名镇、名村和传统村落；（6）历史河湖水系和水文化遗产；（7）山水格局和城址遗存；（8）传统胡同、历史街巷和传统地名；（9）风景名胜、历史名园和古树名木；（10）非物质文化遗产；（11）法律、法规规定的其他保护对象。

通过识别对核心价值体系支撑尚有不足的区域，评估价值载体缺口，结合文物普查和历史文化资源调查，进一步挖掘潜在保护对象，将具备条件的纳入保护名录管理。

（二）明确重点保护内容

突出价值引领，持续开展价值载体系统研究，梳理历史文化资源，明确基于价值体系的北京历史文化遗产重点保护内容。

北京历史文化遗产价值体系与保护重点　　　　　表1

价值主题	价值特征	保护重点
东亚地区早期人地关系的杰出见证	I-1 人类起源	人类起源与农业起源重大遗址
	I-2 农业起源	
	I-3 地质特征及生物多样性	重要地质遗迹景观

续表

价值主题	价值特征	保护重点
见证中华文明赓续的千年古都	Ⅱ-1 地理格局——中华文明统一的多民族国家都城的地理基础	国家地理格局的山川形势
	Ⅱ-2 北京建城史、建都史——中华文明统一的多民族国家发展的伟大见证	北京重要古代城址遗存及金中都遗址、元大都遗址和明清北京城等古代都城遗址
	Ⅱ-3 中国古代都城规制经典范式	以中轴线进行对称布局的明清宫城、皇城、内城、外城四重城市空间与棋盘路网等中国古代后期的城市空间格局特征
	Ⅱ-4 古代礼制建筑群——中华文明礼制文化	与古代国祀、皇祀等祭祀活动相关的坛庙等祭祀建筑
	Ⅱ-5 最具代表性的中国古代皇家建筑	古代宫殿、园林、陵寝、皇家寺观等皇家建筑
	Ⅱ-6 长城体系	与北京长城体系相关的长城边墙、烽火台、关隘、堡寨、城堡、题刻及其他相关设施等遗址遗存
	Ⅱ-7 运河体系	与北京运河体系相关的河道、湖泊、水工设施、桥梁、仓储设施等遗址遗存
引领中国式现代化的首善之都	Ⅲ-1 近现代政权更迭、思想碰撞和革命历程	1840~1949年的政权机构所在地、革命活动相关场所
	Ⅲ-2 新中国首都建设	1949~1978年的体现首都建设的代表性建筑、重大工程
	Ⅲ-3 改革开放后伟大复兴	1978年至今的体现首都发展和"四个中心"建设的代表性建筑、重大工程
多元交流、开放包容的大国之都	Ⅳ-1 多民族文化交流融合	少数民族的生活、生产建筑等遗存
	Ⅳ-2 多元宗教信仰	具有代表性的佛教、道教、伊斯兰教、基督教、天主教和民间传统信仰等各类宗教场所及遗址遗迹
	Ⅳ-3 地方文化传统	具有代表性的胡同、合院、会馆、商肆、集市、茶馆、戏楼、村落、农业、手工业等场所及其历史环境、历史植被等

续表

价值主题	价值特征	保护重点
多元交流、开放包容的大国之都	Ⅳ-4 东西方思想文化与科技交流	具有代表性的近现代行政、外交、展览、金融、医疗、教育、工业、交通等建筑（群）及城市公园
	Ⅳ-5 多元教育思想文化	古代及近现代教育设施和各类名人旧（故）居

（三）动态管理保护名录

强化分级管理。以历史文化价值内涵为核心，加强国家级、市级保护对象名录管理。持续推进世界文化遗产、全国重点文物保护单位、中国传统村落、中国历史文化街区、国家工业遗产、国家级非物质文化遗产代表性项目、中华老字号、中国重要农业文化遗产的申报工作，确保重要价值载体得到重点保护及传承利用。

落实保护名录管理制度，按照《北京历史文化名城保护对象认定与登录工作规程（试行）》，加强预先保护，具有保护价值、符合保护对象认定标准的应及时纳入保护名录。保护对象因不可抗力损毁、灭失或者保护等级、类型发生变化的，应调整保护名录。

二、推进全面系统保护

（一）总体保护要求

1．加强不同时期文化遗产保护

加强对古代保护对象的保护。保护体现古代城市营建、多元民族交流历史的传统遗存及具有历史文化内涵的文化景观。重点加强对反映人类起源、农业起源等文明起源实证载体的保护。

挖掘近现代保护对象的价值。充分挖掘与保护能够反映近现代战争冲突、革命运动与政治体制变革、工商业发展、生活方式变迁、新思想新文化传播、科学技术发展、城市与建筑等方面的历史进程或杰出成就的遗存。重点保护洋务运动、戊戌变法、庚子事变、辛亥革命、五四运动、卢沟桥事变等近现代战争冲突和政治体制变革的遗存纪念地。

进一步保护好体现中国共产党团结带领中国人民不懈奋斗历程的各类保护对象，如反映中国共产党早期革命活动、建立抗日革命根据地、建立新中国等方面伟大历史贡献的遗存。重点保护北大红楼及其周边区域，长辛店"二七"大罢工旧址，平西、平北和冀东抗日根据地，北京香山革命纪念地等遗存，建设相关纪念场所，做好精神传承。

加强对北京能够展现新中国成立、建设和改革开放等重大历史事件的各类保护对象的保护，讲好新中国的北京故事。对体现新中国政治制度建立和发展、体现社会主义制度优越性的保护对象，以及承载城市建设、科技创新、社会发展、对外交往等领域成就的保护对象，进一步挖掘其价值特色。

2．加强区域文化遗产的系统保护

结合长城、大运河国家文化公园建设，永定河、京张铁路、京西古道等线性遗产保护，促进区域历史文化遗产的系统保护。

加强对历史文化资源富集地区的整体保护。保护片区内各类保护对象及其赋存环境，充分挖掘各要素之间的关联性，通过建设文化节点、构建文化线路强化文化遗产的统筹和串联。

3．加强遗存本体与自然人文环境的整体保护

整体保护太行山、燕山及永定河、潮白河、拒马河、北运河、沟

河等主要河流及沟峪。强化山区生态治理，统筹生态涵养与遗产保护，防止水土流失、地面沉降、地震和泥石流等自然灾害对遗产的破坏。不随意改变或侵占河湖水系，提升滨水环境。凸显遗产赋存环境的价值内涵，已纳入保护范围及建设控制地带的山体、河流、农田、林地等应加大保护管控力度，保持各类保护对象所处自然环境特征与地域文化特色，划定必要的景观视廊进行建设控制。

整体保护城镇村自然环境、格局肌理、传统建筑、历史环境要素等物质文化遗存和民间艺术、乡风民俗等非物质文化遗存及其依存的文化生态等，维护历史风貌的完整性、社会生活的延续性、城乡功能的多样性，保持历史文化底蕴，让各时代的历史痕迹和生活记忆都能得到相应保护。

（二）分类保护要点

以彰显保护对象核心价值为目标，突出"价值—载体—环境"的整体保护和系统保护，针对各类保护对象特征制定差异化的保护措施。

（1）**世界文化遗产。**严格遵守《保护世界文化和自然遗产公约》要求，保护世界文化遗产的真实性和完整性，进一步强化对突出普遍价值的阐释展示。落实遗产专项保护（管理）规划，加强保护管理。筛选潜在的申遗项目，以申遗为抓手，促进历史文化遗产保护，带动区域保护更新。

（2）**文物。**严格落实文物保护相关法律法规要求。贯彻保护为主、抢救第一、合理利用、加强管理的方针。加强文物的预防性保护与常态化监测管理，重点治理文物的不合理使用和周边环境建设的不协调等问题。积极推动被不合理使用文物的腾退与合理利用，整治文物建设控制地带内的违规建设行为。贯彻落实保护第一、加强管理、挖掘价值、有效利用、让文物活起来的工作要求。鼓励文物保护与价值阐释展示、公益性再利用、非物质文化遗产传承等活动有效结合。

（3）**历史建筑**。依据相关保护规范，在满足不改变核心价值特征等保护要求的前提下，加强历史建筑保护修缮，制定修缮技术标准和资金补助政策，促进活化利用。

（4）**革命史迹**。突出重大历史事件、历史人物，维护其本体安全和特有的历史风貌。组织开展纪念活动，注重发掘整理、宣传展陈史料和英烈史迹。鼓励将革命史迹与其他保护对象整合，拓展展示路线和内容，形成主题线路。

（5）**历史文化街区**。推动历史文化街区保护规划编制工作。重点保护街区的历史格局、街巷肌理、空间尺度和景观环境，分类保护街区内建筑，逐步整治不协调建筑和景观，延续历史风貌。保持生活的延续性，传承传统文化习俗，维系街区内的生活方式和社区环境。在满足保护要求的前提下，提升基础设施和公共服务设施，改善街区生活条件。

（6）**特色地区**。延续特色地区的主体功能和特色风貌，保护能体现核心价值的历史文化遗产和环境要素，保护反映城市特定发展阶段的既有建筑，增强文化展示功能。针对以生活居住功能为主的特色地区，保护其街巷肌理和传统风貌，保护承载居民记忆和群体情感的历史环境要素，优化提升基础设施，改善居民居住环境；针对以经济产业功能为主的特色地区，保留其核心生产空间布局，推动周边环境治理，适当补充文化展示功能，根据现状产业情况采取产业延续、升级、置换等差异化措施；针对以科教文化功能为主的特色地区，保护其建设空间格局和传统风貌、历史环境、园林景观，传承人文精神。

（7）**地下文物埋藏区**。严格按照《北京市地下文物保护管理办法》的要求进行保护。探索地下文物埋藏区分级管理，重点地区的开发建设在立项前应谨慎论证，对可能存在的有重要价值的古遗址、古墓葬，应事先明确保护方案，全面做好地下文物保护，针对一般地区、重点监测区域及地下文物埋藏区以外的其他区域应坚持"先考古、后出让"，根据考古工作成果确定开发建设方案。

（8）**历史文化名镇、名村和传统村落**。推动历史文化名镇、名村保护规划和传统村落保护发展规划、传统村落集中连片保护利用规划编制。保护镇村选址建设的地形地貌、山水环境、历史格局等整体空间形态，保护凸显价值特色的历史遗存、历史环境要素和特色民居，保护和延续镇村所依托的农田、林地、坑塘等农业要素，保护传统生产生活方式、传统民俗和非物质文化遗产。坚持以用促保，活化利用传统建筑，推进传统民居宜居性改造，改善基础设施和公共服务设施短板，提升环境品质。挖掘传承乡风民俗等传统文化，发展文化旅游等相关产业。加强安全体系建设，完善消防、防灾避灾等安全设施。

（9）**历史河湖水系和水文化遗产**。保护历史河湖水系的总体走向，尽可能维护河道原有形态和传统堤岸，逐步恢复对城市发展具有重要价值的历史河道。合理控制桥、坝、闸等水文化遗产的使用强度。完善水利博物馆、展览馆等，加强水文化的现场展示。

（10）**城址遗存**。完整保护古城墙、古城门和护城河等古城历史轮廓，保护古城道路肌理和重要轴线，保护重要节点遗存。

（11）**传统胡同、历史街巷和传统地名**。加强传统胡同与历史街巷保护，将其中风貌与尺度尚存的传统胡同、历史街巷作为永不拓宽道路加以重点保护，维护好传统街巷肌理与空间尺度。加强地名文化遗产传承与管理，在街区保护更新过程中保留或优先使用传统地名。

（12）**风景名胜**。严格落实《风景名胜区条例》，保护风景名胜的自然环境和文化景观，开展健康有益的游览观光和文化娱乐活动，普及历史文化和科学知识。

（13）**历史名园和古树名木**。最大程度地保留历史名园的空间格局与历史风貌，保护山石水体、古树名木等历史环境要素。严格落实《北京市古树名木保护管理条例》及其实施办法，保护古树名木及其生境，加强古树名木历史、科学、文化等价值的宣传。

（14）**非物质文化遗产**。严格落实非物质文化遗产保护相关法律法规，以"见人、见物、见生活"为原则，保护传承非物质文化遗产及其依存的人文和自然环境。加强非物质文化遗产传承人培养培训。将非物质文化遗产融入物质文化遗产，合理布局非物质文化遗产宣传、展示、传承体验空间。

（15）**老字号**。加强老字号遗存保护，将拥有重要价值文物的老字号企业和符合条件的老字号经营场所原址，依法核定公布为不可移动文物。强化老字号原址原貌保护，涉及搬迁的需征求业务主管部门意见，确需搬迁的按照"拆一还一"或不低于原产权单位评估价值的原则予以补偿。

（16）**农业文化遗产**。挖掘更多重要农业文化遗产，保护并延续其传统的生产、生活方式及衍生文化习俗，划定核心保护范围。将生态环境保护、村庄建设与农业文化遗产保护相结合。

三、统筹协调规划管控

严格落实各类历史文化遗产保护管控要求，汇总保护控制范围。

强化国土空间规划与保护传承要求的管理衔接，妥善处理历史文化遗产保护控制范围与"三区三线"、历史文化保护线的关系，并纳入国土空间规划"一张图"，加强规划管控。

城镇开发边界内防止大拆大建破坏各类保护对象本体及其环境，明确避免集中建设对历史文化遗产本体及其环境造成负面影响的控制性要求和指导性措施，统筹好文化遗产保护和城乡建设发展。

鼓励永久基本农田与各类保护对象保护利用协同。针对大遗址和地下文物埋藏区内的永久基本农田，暂无考古计划的可以在不破坏遗址的前提下有序开展农业活动，有考古计划的应在考古工作开展前办理考古临时用地审批。因保护展示等需要将农用地转为建设用地的，应按规定办理建设用地审批手续。加强高标准农田建设等有计划的农业生产活动中的文物保护；计划开展的农业生产活动选址，应尽可能避开不可移动文物。

生态保护红线内加强与历史文化遗产保护相适应的地域生态环境保护修复，制定针对生态修复工程的指导性措施。在不对生态功能造成破坏的前提下，允许在生态保护红线内、自然保护地核心保护区外，开展依法批准的考古工作、文物等保护对象保护活动以及适度的参观旅游和相关必要的公共设施建设，促进文化和自然遗产的合理利用。

四、明确保护传承空间格局

（一）确立历史文化遗产总体空间格局特征

规划立足北京历史脉络与城市空间形态演化特征，以历史文化价值为导向，以自然地理格局和多元地域文化特色为本底，以重要文化廊道和线路为纽带，以历史文化遗存分布特征为依托，构建"两山一湾、两区两轴、三带多片"的北京城乡历史文化遗产保护传承空间格局。

"两山一湾"指太行山、燕山和北京湾，是北京历史变迁、文化发展的自然本底，是北京的人类之源、历史之根和立都之本。

"两区两轴"指老城、三山五园地区、中轴线及其延长线、长安街及其延长线。其中，老城和三山五园地区是中华传统文化荟萃凝聚、中国式现代化发展集中见证的两大重点区域。中轴线、长安街横亘古今，是古都与首都城市发展的直接实证。

"三带多片"指长城、大运河、西山永定河文化带，及三条文化带串联的若干保护传承重点片区，是北京历史文化遗产及价值较为集中的文化精华区域。

（二）明确不同特征空间保护要求

1．整体保护太行山、燕山自然山水格局

保护好太行山、燕山的地质遗迹和山水环境。开展生态环境评估，探索基于历史文化保护的生态补偿机制，加强生态修复和环境治理，保护生物多样性，保障生态环境安全。持续大尺度造林绿化，恢复历史文化景观。重点保护自然保护区、风景名胜区、森林公园、地质公园、湿地公园等自然保护地。

传承好太行山、燕山的精神文脉。系统梳理、深入挖掘其丰富的文化内涵，讲好中国传统文化中关于人与自然和谐共生思想的精髓。

2．整体保护老城

北京老城是中国都市计划的无比杰作，是中华文明历史、中国近

现代历史进程、中国共产党团结带领中国人民不懈奋斗光辉历程、中华人民共和国成立与发展历程、改革开放和社会主义现代化建设伟大征程的重要见证。严格落实老城不能再拆的要求，坚持"保"字当头，以更加积极的态度、更加科学的手段实施老城整体保护。

（1）整体保护老城的传统空间规划格局。保护传统中轴线；保护明清北京城"凸"字形城廓和宫城、皇城、内城、外城四重城廓；整体保护明清皇城；保护历史河湖水系；保护棋盘式道路网骨架和街巷胡同格局；保护胡同—四合院传统建筑形态；保护平缓开阔的传统城市空间形态；保护景观视廊和街道对景；保护传统建筑色彩和形态特征；保护古树名木和大树。

（2）保护老城的古代礼制建筑群和皇家建筑群。

（3）保护老城多元包容的地方文化代表性遗存。

（4）保护反映从传统社会都城向现代共和国首都转型的近现代代表性遗存。

（5）保护中国共产党领导下开展新中国首都建设和改革开放经济转型以及新时代首都发展伟大成就的代表性遗存。

3．整体保护三山五园地区

三山五园地区是传统历史文化与新兴文化交融的复合型地区，系统保护历史文化遗产和自然景观，恢复山水田园的自然风貌，构建历史文脉与生态环境交融、传统文化与现代文明辉映的整体空间结构。

三山五园是对位于北京西北郊、以清代皇家园林为代表的各历史时期文化遗产的统称。"三山"指香山、玉泉山、万寿山，"五园"指静宜园、静明园、颐和园、圆明园、畅春园。

（1）整体保护三山五园地区山水相依的景观格局；

（2）保护以皇家园林为核心的古典园林；

（3）保护老城与三山五园地区之间水脉相连的水系格局、御道及沿线古村落、京西稻作文化系统；

（4）保护反映近现代政权更迭、思想碰撞和革命历程的重要遗存；

（5）保护反映新中国成立以来教育、科学和文化事业发展的代表性建筑。

4．统筹"两轴"保护发展

"两轴"（中轴线及其延长线、长安街及其延长线）是北京城市空间与功能组织的统领。传统中轴线作为体现中国理想都城秩序的杰作，是国家礼仪制度的重要承载区域。长安街见证了中华民族伟大复兴的历史进程，是大国首都现代化建设发展的重要窗口。

（1）继承发展中轴线和长安街形成的两轴格局。保护中轴线礼序中正、轴线对称的空间秩序和长安街庄严有序、厚重大气的空间形态。

（2）保护中轴线及其延长线上体现国家礼制秩序的重要遗存。

（3）保护长安街及其延长线上体现首都政治中心战略定位的代表性建筑。

（4）保护"两轴"沿线新中国成立以后从首都建设到首都发展、以中国式现代化推进中华民族伟大复兴进程中的代表性建筑。

5．统筹长城、大运河和西山永定河文化带保护发展

长城和大运河是中国古代大尺度自然地理单元人地关系的典型代表，呈现出无与伦比的自然与人文景观高度统一的特征，是中华民族的代表性符号和中华文明的重要象征。永定河是北京的母亲河，是北京城市发展变迁的根脉。以大运河、长城和西山永定河为核心建设三条文化带，是北京历史文化遗产保护传承的重要空间载体和文化纽带。

长城文化带涉及门头沟、昌平、延庆、怀柔、密云、平谷六区；大运河文化带涉及昌平、海淀、西城、东城、朝阳、顺义、通州七区；西山永定河文化带涉及延庆、昌平、海淀、石景山、丰台、门头沟、房山、大兴八区。

（1）保护长城及其赋存环境，包括山川河谷、地质遗迹、防御设施、古村落、古道、民俗非遗、文化景观等自然和文化遗产。

（2）保护大运河及沿线水利工程、古村落、民俗非遗、文化景观等文化遗产。

（3）保护西山永定河沿线的地质遗迹、古道、古遗址、古村落、古建筑、民俗非遗、文化景观等自然和文化遗产。

（4）保护近现代以来与长城、大运河、西山永定河有关的反映革命历程、新中国首都建设和改革开放后伟大成就的代表性建筑（群）、重大工程。

（三）统筹保护传承重点片区保护发展

1．各片区核心价值及保护重点

以各类历史文化遗产的空间聚集性和文化关联性为基础，强化历史文化价值载体的空间分布特征，在老城和三山五园地区以外区域形成十四处特色鲜明、错位互补的集中保护传承重点片区，以历史文化价值为导向进一步完善城市历史文化整体保护格局。

重点保护传承片区指引表　　　　　　表2

分区	重点片区	核心价值	保护重点
朝阳区	奥林匹克中心区—元大都	Ⅱ-2 北京建城史、建都史——中华文明统一的多民族国家发展的伟大见证 Ⅲ-3 改革开放后伟大复兴	1. 保护元大都城墙遗址等古代都城营建的重要物质载体及其周边环境； 2. 保护国家奥林匹克体育中心、国家体育馆、国家游泳中心等承载现代体育、国际交往功能的重要场所
朝阳区	望京—三里屯—北京商务中心区	Ⅲ-2 新中国首都建设 Ⅲ-3 改革开放后伟大复兴	1. 保护原北京电子管厂历史建筑群、原北京有线电厂办公楼等承载近现代工业发展的重要场所； 2. 对三里屯第二使馆区、亮马桥第三使馆区、中国国际贸易中心等承载了现代外交事业与商贸功能的片区进行核心场所挖掘并保护
丰台区	莲花池—金中都	Ⅱ-2 北京建城史、建都史——中华文明统一的多民族国家发展的伟大见证	保护金中都水关遗址、金中都城遗址、莲花池等古代都城营建的重要物质载体及其周边环境
丰台区	卢沟桥—长辛店	Ⅱ-1 地理格局——中华文明统一的多民族国家都城的地理基础 Ⅲ-1 近现代政权更迭、思想碰撞和革命历程 Ⅲ-2 新中国首都建设 Ⅳ-3 地方文化传统 Ⅳ-4 东西方思想文化与科技交流	1. 保护卢沟桥、宛平城及其历史环境； 2. 保护长辛店"二七"大罢工旧址等新民主主义革命的重要场所； 3. 保护北京"二七"机车厂等近现代工业遗产； 4. 保护长辛店原冯家大院、长辛店聚来永副食店、长辛店原第一理发店等反映老镇居民传统生活特征的重要场所
石景山区	模式口—八大处—首钢	Ⅱ-1 地理格局——中华文明统一的多民族国家都城的地理基础 Ⅲ-1 近现代政权更迭、思想碰撞和革命历程 Ⅲ-2 新中国首都建设 Ⅲ-3 改革开放后伟大复兴 Ⅳ-2 多元宗教信仰 Ⅳ-3 地方文化传统	1. 保护慈善寺、显应寺等宗教场所遗存； 2. 保护模式口历史文化街区等承载传统城乡居民关系的重要场所； 3. 保护八宝山革命公墓重要国家纪念地； 4. 保护首钢等承载近现代工业发展的重要场所及冬奥遗产

续表

分区	重点片区	核心价值	保护重点
门头沟区	沿河城—爨底下—东胡林人遗址	Ⅰ-2 农业起源 Ⅰ-3 地质特征及生物多样性 Ⅱ-1 地理格局——中华文明统一的多民族国家都城的地理基础 Ⅱ-6 长城体系 Ⅲ-1 近现代政权更迭、思想碰撞和革命历程 Ⅳ-3 地方文化传统	1. 保护东胡林人遗址等新石器时代遗址； 2. 保护永定河及沿线湿地公园、森林公园； 3. 保护古香道、古军道、古商道等京西古道及其沿线交通遗存； 4. 保护沿河城等长城防御体系的重要物质载体及其周边环境； 5. 保护八路军冀热察挺进军司令部旧址等革命史迹； 6. 保护爨底下村等传统村落
房山区	金陵—周口店—云居寺—琉璃河遗址	Ⅰ-1 人类起源 Ⅰ-3 地质特征及生物多样性 Ⅱ-1 地理格局——中华文明统一的多民族国家都城的地理基础 Ⅱ-2 北京建城史、建都史——中华文明统一的多民族国家发展的伟大见证 Ⅱ-5 最具代表性的中国古代皇家建筑 Ⅳ-2 多元宗教信仰 Ⅳ-3 地方文化传统	1. 保护周口店遗址、镇江营遗址等新旧石器时代的重要物质载体及其周边环境； 2. 保护上方山国家森林公园等自然保护地； 3. 保护琉璃河大桥等重要交通遗存； 4. 保护琉璃河遗址、窦店遗址等重要古城遗址； 5. 保护金陵、庄亲王园寝、奕绘、顾太清庄园及园寝等古代皇家陵寝、贵族墓葬及其周边环境； 6. 保护房山云居寺塔及石经、万佛堂、孔水洞石刻及塔、十字寺遗址等宗教场所或历史文化遗存； 7. 保护南窖村、水峪村等传统村落

续表

分区	重点片区	核心价值	保护重点
通州区	城市副中心	Ⅱ-1 地理格局——中华文明统一的多民族国家都城的地理基础 Ⅱ-2 北京建城史、建都史——中华文明统一的多民族国家发展的伟大见证 Ⅱ-7 运河体系 Ⅲ-1 近现代政权更迭、思想碰撞和革命历程 Ⅲ-3 改革开放后伟大复兴 Ⅳ-2 多元宗教信仰 Ⅳ-3 地方文化传统 Ⅳ-4 东西方思想文化与科技交流	1. 保护路县故城遗址、通州古城、潞县古城等古代区域行政建制变迁的重要物质载体及其周边环境； 2. 保护大运河及沿线水工设施等运河遗存； 3. 保护平津战役指挥部旧址等革命史迹； 4. 保护北京艺术中心、北京城市图书馆、北京大运河博物馆、城市绿心等城市副中心规划建设的重大工程； 5. 保护通州燃灯塔、紫清宫等宗教传播与发展的重要场所； 6. 保护十八个半截胡同等重点区域； 7. 保护通州近代学校建筑群等近现代建筑遗存
昌平区	居庸关—十三陵—银山塔林—白浮泉遗址	Ⅱ-5 最具代表性的中国古代皇家建筑 Ⅱ-6 长城体系 Ⅱ-7 运河体系 Ⅲ-1 近现代政权更迭、思想碰撞和革命历程 Ⅲ-2 新中国首都建设 Ⅳ-2 多元宗教信仰 Ⅳ-3 地方文化传统	1. 保护十三陵等皇家陵寝及其周边环境； 2. 保护居庸关等长城防御体系的重要物质载体及其周边环境； 3. 保护京张铁路南口段等近代铁路遗产； 4. 保护大运河（白浮泉遗址）等运河遗存； 5. 保护十三陵水库等首都建设的重大工程； 6. 保护银山塔林等宗教场所或历史文化遗存； 7. 保护德陵村、康陵村、茂陵村、万娘坟村等传统村落

续表

分区	重点片区	核心价值	保护重点
大兴区	南苑—团河城行宫遗址	Ⅰ-3 地质特征及生物多样性 Ⅱ-5 最具代表性的中国古代皇家建筑 Ⅲ-1 近现代政权更迭、思想碰撞和革命历程 Ⅲ-2 新中国首都建设 Ⅳ-4 东西方思想文化与科技交流	1. 保护团河城行宫遗址等古代皇家园林的重要物质载体及其周边环境； 2. 保护南苑兵营司令部旧址（位于丰台区内）等革命史迹； 3. 保护南苑机场等近现代建筑遗存
怀柔区	黄花城—慕田峪—怀柔科学城—雁栖湖	Ⅱ-6 长城体系 Ⅲ-3 改革开放后伟大复兴 Ⅳ-1 多民族文化交流融合 Ⅳ-2 多元宗教信仰	1. 保护长城（慕田峪段）等长城防御体系的重要物质载体及其周边环境； 2. 保护雁栖湖国际会都、怀柔科学城等承载现代国际交往与高新技术产业发展功能的特色地区； 3. 保护喇叭沟门满族乡、长哨营满族乡等象征多元民族交流的重要村落； 4. 保护红螺寺等宗教场所或历史文化遗存
平谷区	将军关—上宅遗址	Ⅰ-1 人类起源 Ⅰ-2 农业起源 Ⅰ-3 地质特征及生物多样性 Ⅱ-6 长城体系 Ⅲ-1 近现代政权更迭、思想碰撞和革命历程	1. 保护上宅遗址、马家坟遗址等旧石器和新石器遗址； 2. 保护金海湖风景名胜区、黄松峪国家地质公园等自然保护地； 3. 保护长城（将军关）等长城防御体系的重要物质载体及其周边环境； 4. 保护鱼子山抗战遗址等革命史迹
密云区	古北口—司马台—白马关—密云水库	Ⅰ-3 地质特征及生物多样性 Ⅱ-6 长城体系 Ⅲ-1 近现代政权更迭、思想碰撞和革命历程 Ⅲ-2 新中国首都建设 Ⅳ-3 地方文化传统	1. 保护云蒙山国家地质公园、古北口森林公园等自然保护地； 2. 保护长城（司马台段）等长城防御体系的重要物质载体及其周边环境； 3. 保护古北口战役阵亡将士墓、白乙化烈士陵园等革命纪念地； 4. 保护密云水库等首都建设的重大工程； 5. 保护黄峪口村、白马关村等传统村落

续表

分区	重点片区	核心价值	保护重点
延庆区	八达岭—官厅水库—世园会—冬奥赛区	Ⅰ–3 地质特征及生物多样性 Ⅱ–1 地理格局——中华文明统一的多民族国家都城的地理基础 Ⅱ–6 长城体系 Ⅲ–1 近现代政权更迭、思想碰撞和革命历程 Ⅲ–3 改革开放后伟大复兴 Ⅳ–1 多民族文化交流融合 Ⅳ–4 东西方思想文化与科技交流	1. 保护野鸭湖湿地自然保护区等自然保护地； 2. 保护八达岭长城、岔道城等长城防御体系的重要物质载体及其周边环境； 3. 保护京张铁路八达岭段等近代铁路遗产； 4. 保护延庆古崖居等多元民族生活遗迹； 5. 保护世园会、冬奥会场馆等承载"四个中心"重要功能的重大工程建设

2. 重点片区管控总体策略

强化历史文化价值的阐释展示，坚持在保护中发展、在发展中保护，加强对十四处重点片区的总体管控。

保护历史文化价值载体。挖掘历史文化遗产所承载的核心价值。整体保护重要历史文化遗产的本体及其周边环境，加强风貌管控；保护承载核心价值的物质载体。

强化片区核心价值的阐释展示。深入研究各重点片区，围绕核心价值阐释其历史渊源、发展脉络与基本走向。对能够体现核心价值的物质载体本身，要通过直接展示或配合以辅助手段阐释其内涵与价值；对通过见证重要历史事件或名人而产生价值的场所，要在其功能、空间规划中对核心价值予以回应，并设立标识予以展示说明。

联动片区打造文化主题线路。挖掘各重点片区间的关联性，依托古道、绿道、铁路、河流、沟峪、山脊等线性空间资源，依托核心价值打造各具特色的文化主题线路。强化沿线历史文化遗产统筹保护，选择具有代表性的局部示范段重点展示。

改善片区民生环境。加大重点片区环境综合整治力度，完善基础设施建设，切实改善人居环境。依托重点片区开展多元文化活动，丰富居民生活，提高居民生活品质。

促进片区经济发展。依托重点片区与文化主题线路，加强遗产保护展示和旅游发展相关的基础设施和服务设施建设，促进沿线村落的文化交流和经济产业发展。

第四章

构建传承体系，弘扬中华文化

一、深化价值的挖掘阐释

（一）助力建设中华文明标识体系的北京篇章

全方位、多层次探索中华文明起源及北京融入中华文明多元一体格局的历史进程，深入挖掘古都文化、京味文化。加大跨领域、跨学科创新研究力度，开展太行山、燕山自然山水文化在北京都城形成和统一国家构建中的重要意义研究，进行整体性阐释。围绕考古实证北京人类起源、农业起源、城市营建、多元文明交流互鉴等关键问题，持续开展周口店"北京人"遗址、东胡林人遗址、琉璃河遗址、路县故城遗址、金陵遗址、金中都遗址等大遗址和北京老城、圆明园及长城的考古工作；实施永定河等主要水系及支流的区域考古调查，构建聚落形态变迁框架，厘清北京史前文化演变脉络。持续开展考古测绘、调查、勘探和发掘，搭建考古基础信息数据平台，推动考古研究成果矢量化、数字化。

加强对承载北京历史文化遗产核心价值具有突出意义、重要影响的历史文化遗产的研究，系统挖掘、梳理和提炼其历史故事、文化价值、精神内涵，遴选能够代表国家形象和中华民族独特精神的标识，谱写中华文明标识体系的北京篇章。加强长城、北京故宫、周口店"北京人"遗址、颐和园、天坛、明十三陵、大运河、北京中轴线八处世界文化遗产的保护传承，推荐有条件的项目列入申报世界遗产预备名单。

（二）深入阐发首都社会主义现代化伟大征程

不断丰富拓展革命史迹、历史建筑、特色地区等各类保护对象保护名录，大力传承弘扬红色文化、创新文化。基于中国共产党史、新中国史、改革开放史、社会主义发展史的研究，保护党和人民在各个历史时期奋斗中形成的重要建设成果，梳理总结体现五四精神、"进京赶考"精神、"两弹一星"精神、劳模精神、劳动精神、工匠精神、科学家精神、北京冬奥精神等中国共产党人精神谱系的代表性遗产，发挥其时代价值。

二、加大城乡建设中的文化传承力度

（一）保护传承好古都空间格局

以规划引导城乡历史文化遗产的保护传承，尊重和传承优秀传统规划设计的城市基因，促进城乡历史文化保护与规划建设相结合。

加强对北京古都传统营城理念、城市规划的系统研究，继承弘扬山川拱卫、纲维有序、礼乐交融的壮美空间格局。强化老城和三山五园地区作为首都中枢的重要功能，维护历史格局。延续"两轴"的空间统领作用，在轴线重要节点预留重要国家文化功能和对外交往功能。突出大尺度山水格局，整体营造"山—水—城"的融合秩序，凸显山水环境在城乡建设中的重要作用。

（二）培育具有文化特色的城市空间

加强城市设计引导，彰显地域风貌特色。构建城市空间秩序体系，强化建筑风貌管控，统筹协调建筑高度、体量、色彩与第五立面等各项立体空间要素。

推动公共环境艺术发展，构建与历史底蕴相契合的人文环境。充分挖掘公共空间背后的历史信息与文化内涵，提升公共空间设计水平，融入文化要素，充分展示人文内涵。加强城市文化基因体系建设，强化城市视觉识别系统，彰显地域特色。

促进当代建筑文化的承古开新，创作根植于历史环境的现代建筑。汲取传统建筑的营建智慧，振兴传统建筑文化，推进其现代传承应用，融通中外，提升当代建筑设计理论与方法水平，建设无愧于历史与时代的民族建筑精品，呈现与历史格局秩序和谐共生的时代风韵。加大文化传承重点地区建筑设计的审核力度。

加强对乡村地区自然环境景观、村庄风貌管控。鼓励采用具有地方特色的建筑材料、建造方式开展村庄建设，保护传承有文脉特色的文化要素、建筑特色和构建特点，更好地体现地域特征。强化乡村与自然山水的景观联系，塑造有文化内涵的村庄空间环境。

（三）强化文化传承的更新改造

切实加强城市更新中的历史文化遗产保护，坚持以保留保护为主、留改拆并举，老街区、老居住区、老镇、老村、老厂区等更新改造前，预先进行历史文化价值评估，及时认定、公布保护对象，落实保护措施。

落实《北京市城市更新条例》，传承延续传统肌理，留住城市更新区域内特有的地域环境、文化特色、建筑风格等"基因"，禁止破坏地形地貌、损害和砍伐古树名木等改变历史格局和环境风貌的行为。在城市更新中严禁大拆大建、拆真建假、以假乱真，不大规模、成片集中拆除现状建筑。不随意拆除具有保护价值的老建筑、老民居、老牌坊、古井、古桥等建（构）筑物，切实保护能够体现城市特定发展阶段、反映重要历史事件、凝聚社会公众情感记忆的既有建筑、空间场所和标志物。新建改建项目应延续传统肌理，展现城市特色。

三、加强文化遗产的保护传承

（一）构建以价值为核心的保护利用体系

文化遗产的保护利用应以彰显历史文化价值为核心，做好价值阐释，传承文化精神。针对不同的保护对象采用更加开放灵活的保护利用模式，着力解决保护利用不平衡、不充分的问题。

保护利用必须以确保保护对象本体安全为前提，不得擅自破坏、损害保护对象，影响环境风貌，应符合遗产价值，优先服务于公共公益功能。

（二）加大建筑类文化遗产的开放力度

坚持以用促保，充分提升古建筑、近现代重要代表性建筑等文物建筑、历史建筑、历史名园等建筑类文化遗产的保护利用水平，做到不求所有、但求所保，向社会开放。积极推进现状不合理使用的重点文物、历史建筑腾退保护，实现全面开放或者适度开放。鼓励非国有文物、历史建筑、历史名园主动向公众开放、提供展览展示服务，相关主管部门应予以指导和帮助。

尊重文物、历史建筑、历史名园等历史功能，根据《北京市文物建筑开放利用导则（试行）》《文物保护利用规范　名人故居》《革命旧址展示导则》《北京工业遗产管理办法（试行）》等相关规定，结合其历史文化价值特征，用作国事活动、公益办公、文化展示、旅游休闲等场所。

鼓励探索功能置换、兼容使用、租金优惠、资金补贴等政策创新，结合需求引导功能转换，促进保护与利用相统一。因保护传承、价值阐释以及完善城市功能、补齐城市短板等需求，可依照法律法规要求履行相应报批程序后，转换文物建筑、历史建筑、历史名园等使用用途，转换后的用途需符合保护对象的保护要求和价值特征。

鼓励以文化IP（知识产权）、城市品牌为抓手，促进集群式保护利用，丰富文化业态，延展价值链条。

（三）改善街区类文化遗产的人居环境

因地制宜改善提升历史文化街区和历史文化名镇、名村、传统村落基础设施和公共服务水平，加强防灾减灾、环卫设施建设，提高人居环境质量，建立以居民为主体的保护实施机制。推动老城平房区申请式退租（换租）、保护性修缮、恢复性修建，拆除院内违法建筑，恢复传统四合院基本格局。在有效保护的基础上改善居住环境，合理高效利用腾退房屋，推进平房住宅成套化改造，增设适老化设施和厨卫设施，让历史文化和现代生活融为一体，恢复具有老北京味的生活场景。探索建立适合平房区实际的物业服务模式，推进平房院落自主更新。

多措并举探索历史文化街区和历史文化名镇、名村、传统村落功能更新的新路径。创新平房区存量更新机制，促进风貌织补与资产利用协同，加强社区共建与文化织补协同，提升历史文化街区保护更新，培育"小而精""小而特"的文化街巷（坊），丰富文化和旅游消费场景，拓展城市消费新空间，带动区域复兴。鼓励传统村落执行更开放的产业发展政策，支持名镇、名村、传统村落发展壮大集体经济，探索农村集体经济组织和村民依法以集体经营性建设用地使用权、闲置宅基地、房屋等入股方式参与传统村落保护发展。积极推进门头沟、密云、房山、昌平、延庆、怀柔、平谷传统村落集中连片保护发展，实现留住乡亲、护住乡土、记住乡愁。加强传统农耕文化与传统村落保护相结合，积极推动农业文化遗产项目所在地申报传统村落。

（四）加强遗址类文化遗产的价值阐释

挖掘古遗址类文物、地下文物埋藏区、城址遗存等遗址类保护对象的文化内涵。综合评估全市重要古人类遗址、历史城址、大型建筑群和园林遗址、陵寝和大型墓葬群、红色遗址的价值，围绕突出国家属性，突出政治中心、文化中心和国际交往中心建设，开展综合性的保护传承利用，打造国家级、市级纪念地，建设遗址公园、博物馆、纪念馆等公共文化设施。

聚焦考古遗址公园建设，整体谋划全市考古遗址公园布局，提升周口店和圆明园等国家考古遗址公园水平，重点建设琉璃河国家考古遗址公园和路县故城考古遗址公园，持续推进上宅、东胡林人、金中都、金陵等遗址公园建设。聚焦城市考古，系统开展蓟、唐、辽、金、元等北京历代城址考古及保护展示，结合长城、大运河、中轴线、圆明园等考古工作，加强遗址遗迹与城市公共空间建设的主题文化阐释。聚焦红色革命遗址，围绕重大时间、重大节点，研究确定一批重要标识地，建设中国共产党人的精神殿堂、中国人民的精神家园、中华民族的精神高地。

（五）振兴非物质文化遗产和老字号

加强非物质文化遗产的系统性保护，融入生产生活，保障传承空间。针对传统技艺、传统美术、传统医药等蕴含精湛工艺的非物质文化遗产，依托非物质文化遗产场所打造集工艺展示、产品产出于一体的传承基地。针对传统戏剧、传统舞蹈、传统体育、游艺与杂技等各类演艺形式，结合文物、历史建筑、历史名园等文化空间进行沉浸式展演。针对各村落在发展过程中衍生出的富有当地特色的民俗和非物质文化遗产，融入村庄发展，助力乡村振兴。推动非物质文化遗产与旅游融合，丰富北京文旅资源。

加强老字号原址、原貌保护。积极落实《进一步促进北京老字号创新发展的行动方案（2023—2025年）》，鼓励老字号以拓宽传承载体的利用方式，开设博物馆、展览馆、体验馆和文化馆等展示体验空间。提质升级老字号特色产品与服务，擦亮金字招牌，激发内生活力。

（六）助力城乡可持续发展

突出文化遗产保护利用在弘扬价值、环境整治、社会参与、经济发展、宣传教育等方面的重要作用，加强文化遗产保护与周边城乡建设的协同联动，优化城乡空间布局，完善功能，提升活力，加强城市特色风貌塑造和城市生态修复，延续城市历史文脉，助力城乡可持续发展。提升大遗址、历史文化街区、线性遗产等大型文化遗产优化区域空间格局的作用，以文化为纽带，串联文化、生态、产业、城市、乡村，带动区域经济发展转型升级。

（七）支撑国家重大战略

推动国家文化公园建设。以国家文化公园建设为抓手，优化城乡功能布局、促进沿线区域发展，构建历史文化遗产连片、成线的整体保护格局，塑造融合历史文脉、生态环境和现代设施的城乡景观。

推动国家"文化重器"落户北京。紧紧围绕新时代国家文化发展需求，推动具有国家代表性、辐射作用强、有突出意义和重要影响力或能发挥对外文化交流重要作用的现代重大公共文化设施建设。

深入落实爱国主义教育实施方案。聚焦建党、建军、建国等重大时间节点及党和国家中心任务，依托红色历史文化遗产，组织开展系列纪念活动和群众性主题教育。

四、促进历史文化的创新弘扬

（一）提升历史文化展示水平

依托重要的自然和人文景观资源，以价值主题为主线，深入挖掘历史故事，以各类历史文化资源为主干，以风景名胜、自然景观、文化场馆等为补充，利用运河、长城、铁路、古道等线性遗产，有机关联、串珠成链，建设文化主题线路。

与长城和大运河国家文化公园建设、中轴线保护传承、长安街环境整治提升、三山五园国家文物保护利用示范区建设、红色文化保护传承等相衔接，建设一批具有中华文化标识的重大文化主题线路，扩大首都文化的国际影响力。鼓励各区以历史道路、山川河流以及历史遗迹、历史事件、历史人物为依托，凝练塑造区级文化主题线路，展示地域文化特色。

坚持价值导向，以价值阐释为重点建设文化主题线路，揭示其背后蕴含的哲学思想、人文精神、价值理念，凸显真实性、公益性，避免过度商业化、娱乐化。丰富文化主题线路的利用方式，鼓励创新开展研学教育、徒步骑行、自驾露营、民俗展演、体育赛事等特色活动。积极推动文化主题线路与旅游产业融合发展，推进生态旅游、精品民宿、森林康养、休闲农业等业态培育。

（二）强化文化教育体系

持续推进历史文化进校园。将历史文化保护传承全方位融入启蒙教育、基础教育、职业教育、高等教育、继续教育各阶段，推动高等院校、职业学校开设历史文化保护传承相关学科专业建设。依托丰富的北京城乡历史文化遗产，鼓励开展各类历史研学、教育实践等各类活动。

培养北京历史文化遗产保护传承人才队伍。坚持培育人才和挖掘人才相结合，构建北京历史文化遗产保护传承人才高地平台。建立健全非物质文化遗产传承人和传统工匠的培训、评价机制。扶持民间骨干，培养扎根基层的乡土文化能人、民间文化传人等。

（三）服务全国文化中心建设

充分发挥历史文化遗产保护传承在全国文化中心建设中的空间统领作用，全力做好首都文化这篇大文章，促进中华优秀传统文化、革命文化、社会主义先进文化不断结合创新，取得更加丰硕的文化成果。

围绕古都文化、红色文化、京味文化、创新文化，利用好北京文脉底蕴深厚、文化资源集聚的优势，持续推进"博物馆之城""书香京城""演艺之都"建设，打造具有鲜明特色的城市文化品牌。

搭建多种类型、不同层级的文化交流展示平台，吸引全国各地优秀文化在首都集中展示、交流互动。利用保护修缮好的各类会馆、名人旧（故）居等保护对象，为各地文化精品进京提供展示展演空间。

（四）丰富历史文化供给内容和形式

融入生产生活与文艺创作。组织开展传统节庆活动、纪念活动等形式多样的文化主题活动，结合重大历史事件节点设置建城、建都等纪念日，深化社会对优秀文化的认知。

创新宣传方式。充分发挥报纸、书刊、电台、互联网等新闻媒体和新媒体平台作用，鼓励通过新闻报道、电视剧、电视节目、纪录片、动画片、短视频等形式，推动个人、社会组织及企业参与文化推广，拓宽宣传渠道。持续开展历史文化"进机关""进校园""进社区"等活动，不断强化北京历史文化的影响力、凝聚力和感召力。

（五）推动北京历史文化走向世界

建设国家会客厅。对于突出彰显中华优秀传统文化、规模和区位符合相应条件的中华文明标识，预留国家礼仪活动服务功能，建设国家会客厅，推动文明交流互鉴。

深化国际交流合作。深化共建"一带一路"国家交流合作，推进历史文化保护传承经验共享。积极扶持具备条件的历史文化场所举办适当的国际交流活动，做强北京文化论坛、北京国际设计周、中国（北京）国际运河文化节、北京文博会等品牌文化活动，打造有国际影响力的全球顶级文化交流平台。

提升国际传播能力。坚持首都定位、国家站位、全球视野，充分运用海外中国文化中心、孔子学院等平台及国内外博览会等，讲好中国故事、北京故事。持续推进文化"走出去"工作，发挥历史文化在服务国家外交大局中的独特作用，提升北京历史文化遗产的国际影响力。

五、加强京津冀历史文化遗产协同保护

（一）建立协同保护机制

加强顶层设计，协同推进京津冀历史文化遗产保护政策制定和永定河保护等立法工作。探索建立京津冀历史文化遗产保护常态化沟通机制，对涉及跨区域的历史文化遗产政策措施，加强沟通协商。鼓励依托长城、大运河、明清皇家陵寝、燕都遗址、古人类遗址等京津冀重要历史文化遗产建立合作研究平台，提升京津冀历史文化协同研究水平。推动京津冀高校合作，开展历史文化遗产保护传承领域的合作

研究和联合培养。

（二）完善京津冀历史文化遗产保护传承体系

将历史文化遗产保护传承作为推动京津冀协同发展的重要任务和战略支撑。凝练京津冀文化特征，全面展现燕赵大地、京畿重地在中华文明历史、中国共产党史、新中国史、改革开放史、社会主义发展史中的重要作用。坚持整体保护理念，系统推动京津冀历史文化名城、名镇、名村、传统村落和历史文化街区等保护，延续城乡历史文脉。全面保护好世界遗产，建设京津冀中华文明标识，鼓励京津冀开展世界遗产联合申报工作。推动大遗址保护和国家考古遗址公园建设。整体保护燕山、太行山等山水格局，促进历史文化、山水格局与城乡建设相融合。深入实施非物质文化遗产传承发展工程，探索建立文化生态保护区，着力提升非物质文化遗产系统性保护传承水平。保护铁路遗产、矿冶遗产、港口遗产等各类工业遗产和农业文化遗产，留住工农业生产建设的历史记忆。深入挖掘海洋文化资源和遗产，弘扬国家海洋文化。

统筹长城文化带、大运河文化带、太行山文化带、海洋文化带保护发展。落实国家战略，协同建设长城、大运河国家文化公园，联合开展申报太行山国家文化公园的可行性研究，保护好、传承好、利用好长城、大运河、太行山沿线和渤海湾历史文化生态资源，统筹推进文化保护传承、生态保护修复、民生改善提升、文旅融合发展等重点工程，提升文化遗产保护传承利用效能，树立具有民族性、世界性的文化地标。传承弘扬红色文化，深入挖掘五四运动、抗日战争、平津战役、"进京赶考"、新中国成立等主题红色文化资源，打造京津冀经典红色文化线路。

第五章

完善实施机制，保障有效管理

一、健全工作机制

（一）强化统筹协调

切实加强党的领导。建立健全党委领导、政府统筹、单位实施、公众参与、社会监督的历史文化遗产保护传承工作机制。各级党委和政府全面落实党中央、国务院决策部署，充分认识在城乡建设中加强历史文化保护传承的重要意义。加强北京历史文化名城保护委员会（简称为名城委）总体谋划、统筹协调、整体推进和督促落实的职能，完善名城委及其办公室推进历史文化遗产保护传承的工作机制，落实重大事项请示报告相关要求。将历史文化遗产保护传承纳入干部培训课程，提高各级党政领导干部在城乡建设中保护传承历史文化的意识和能力，牢固树立保护也是政绩的观念。

强化上下协同和部门联动。强化全市一盘棋，做好央地协同、区域协同、市区联动，构建综合管理服务体系，加强规划统筹安排、政策制度衔接、资源要素共享。充分发挥属地政府的作用，完善区级保护传承机制建设，强化老城和三山五园地区等重点片区保护传承的统筹协调和组织实施工作。加强历史文化遗产与生态环境协同保护，探索建立多部门协调沟通、联合执法机制。

切实发挥基层政府的保护传承重要作用。推动保护管理重心下移，建立区级统筹、街道主体、部门协作、专业力量支持、社会公众广泛参与的历史文化保护实施机制，充分发挥责任规划师的工作优势，为区级保护工作提供基础支撑。持续加强各级历史文化遗产保护工作队伍建设，提升基层保护工作队伍能力和水平。

（二）加强规划传导

强化规划纵向传导。贯彻落实国家有关要求，规划批复后纳入国土空间规划"一张图"，做好历史文化名城、名镇、名村、传统村落和历史文化街区等相关保护规划及三条文化带保护发展规划等重大专项保护规划的传导衔接，指导各区历史文化保护传承工作。

加强规划横向协同。深化北京历史文化遗产保护传承体系规划与国民经济和社会发展规划等相关规划的衔接。针对历史文化资源富集、空间分布集中的地域，以及非物质文化遗产高度依存的自然环境和历史文化空间，明确区域整体保护利用的空间管控要求。

（三）推动多方参与

探索多种途径开展历史文化遗产保护公众参与。建立广泛的社会参与机制，畅通参与渠道，为企业、高等院校、科研院所、社会组织、志愿者以及热心人士参与历史文化遗产保护传承工作提供平台。鼓励多元主体在历史文化遗产保护传承的规划、建设、管理各环节发挥积极作用。强化保护责任人制度，明确保护对象所有人和使用人的责任义务。进一步优化营商环境，优化审批管理，制定优惠政策，鼓励市场主体持续投入历史文化遗产保护传承工作。

完善专家咨询机制。持续推进专家智库建设，开展历史文化遗产保护传承重点问题专题研究，完善咨询、评审机制，发挥专家决策参谋、业务咨询、工作指导的作用。建立涵盖国土空间规划、历史文化名城保护、文物保护、建筑工程、历史人文地理、考古等不同领域的区级层面历史文化遗产保护传承专家库，完善专家基层工作指导机制。

优化奖励激励机制。研究制定奖补政策，通过以奖代补、资金补助等方式支持城乡历史文化遗产保护传承工作。持续开展历史文化遗产保护传承示范案例征集工作，及时总结经验做法。对在保护传承工作中作出突出贡献的组织和个人，按照国家和北京市有关规定予以表彰或者奖励。

（四）加强监督检查

贯彻落实《关于在城乡建设中加强历史文化保护传承的意见》要求，健全监督检查机制，严格依法行政，加强执法检查，完善评估机制，鼓励社会监督。

加大执法力度，严肃纠治违法行为。街道办事处、乡镇人民政府在区人民政府的领导下，加强对辖区内的历史文化遗产保护情况的巡查，及时处置危害保护对象的行为，并向相关主管部门报告。加大历史文化遗产保护督察力度，推动历史文化遗产保护"党政同责""一岗双责"有效落实，加强监督检查和问责问效。

完善定期评估与考核机制。加强历史文化资源调查评估，及时扩充保护对象，丰富保护名录。落实"一年一体检、五年一评估"制度，充分结合城市总体规划体检评估等工作的开展，进行历史文化名城专项评估，对各有关部门保护管理要求落实情况、各类保护对象保护传承情况等开展检查评估，评估结果作为改进政府服务效能、制定调整政策的重要依据。将历史文化保护传承工作纳入文明城区测评体系。加大城乡历史文化保护传承的公益诉讼力度。

鼓励社会参与监督。加强文化遗产行政信息公开，向公众提供历史文化遗产保护传承信息查询服务，接受社会监督。鼓励公民、法人和其他组织举报涉及历史文化保护传承的违法违规行为。

二、完善保障机制

（一）健全配套政策体系

坚持全要素、全流程管理，结合国家关于历史文化保护传承的最新要求，筑牢法治保障，完善保护管理配套政策体系，推进保护利用的全过程管理，优化对各类保护对象实施保护、修缮、改造、迁移、利用的审批管理，加强事中事后监管。

适时推动历史文化遗产保护领域地方性法规修改工作。探索多种途径实施文物腾退、保护利用和开放展示。推进传统村落保护管理政策制定。完善申请式退租（换租）、保护性修缮和恢复性修建等配套政策机制。探索保护利用底线管理模式，分类型、分区域建立项目准入正负面清单，定期评估，动态调整。加强政策协同，研究制定规划、消防、环保、证照办理等方面更管用、更好用、更具操作性的利用支持政策。建立全生命周期的建筑管理制度，结合工程建设项目审批制度改革，加强对既有建筑改建、拆除管理。

（二）加强资金保障

健全历史文化遗产保护传承工作的财政保障机制，用好全市历史文化保护资金，重点支持国家级、市级以及突出中华文明和北京历史文化遗产核心价值的项目。规范保护资金转移支付管理，增强各区保护工作的自主权和积极性，推动权责匹配落实到位。加强资金成本绩效管理，有效发挥财政资金的保障作用。用足用好国家各项支持政策，积极发挥中央财政专项资金的引导作用。探索历史文化名镇、名村、传统村落、历史建筑保护资金修缮补助机制。

引导社会资本参与历史文化遗产保护工作。鼓励按照市场化原则加大金融支持力度，充分利用市场化手段和市场化运作方式，有效拓宽资金渠道，逐步形成政府、社会、居民等多方出资，分工合作，共同参与保护的新格局。

（三）推进数字化赋能

加快推进历史文化保护与管理信息平台建设，完善各类保护对象数字化技术标准，开展数字化信息采集和档案建立，增强数字化展示互动，逐步实现数字化管理。加强数据多部门整合共享，建立历史文化遗产保护监测指标体系，动态采集、接入汇集多源数据，用好国土空间规划"一张图"，实现历史文化遗产保护与利用的实时管控、动态监测、定期评估和及时预警。利用卫星遥感技术监测历史文化遗产保护控制范围内的建设行为。搭建公众参与和历史文化遗产信息公开运营平台，切实推动空间治理体系和治理能力现代化。利用现代科技手段对历史文化进行数字化保护、加工、展示，试点智慧技术应用场景，创新历史文化内容传播方式。

（四）建立安全应急管理机制

强化安全保障机制。坚持预防为主，加强日常维护，建立健全监测预警机制。加强历史文化遗产安全风险评估检查，提升安全防范工程水平和抗风险能力，对有损毁风险的，应及时加固修缮，消除安全隐患。完善综合防灾体系，加强消防安全、消防力量建设，推动抗震防震等防灾减灾工程建设，综合运用人防、物防、技防等手段，提高各类历史文化遗产的防灾减灾救灾能力。

探索建立健全应急管理机制。建立针对遗产安全等突发事件和考古新发现的联合指挥处置机制，重点抓好突发事件信息共享、紧急敏感事件会商评估等工作，加强专家指导和支持，提升历史文化遗产安全保障水平。

三、加强项目管理

进一步加强历史文化遗产保护传承项目库建设，以突出历史文化遗产核心价值为重点，提前研究储备重点项目，按照成熟一批、实施一批的原则，将具备可实施性的项目纳入预算安排。

积极推进试点示范项目机制。围绕政策堵点、实施难点、问题痛点开展试点示范项目，建立试点示范沟通协调机制，加强对试点示范项目的支持、指导和监督，综合运用政策、标准、项目配套等资源，促进试点示范项目经验推广应用。

Chapter 1

General Principles

I   Guiding Ideology

Upholding Xi Jinping Thought on Socialism with Chinese Characteristics for a New Era as the guiding principle, fully implementing the spirit of the 20th National Congress of the CPC and the 2nd and 3rd Plenary Sessions of the 20th CPC Central Committee, thoroughly studying and implementing Xi Jinping Thought on Culture, putting into practice the important speeches and instructions concerning Beijing given by General Secretary Xi Jinping, building stronger cultural confidence, upholding fundamental principles and breaking new ground, focusing on highlighting the historical and cultural value of Chinese civilization, reflecting the spiritual aspirations of the Chinese nation and presenting a comprehensive and authentic portrayal of both ancient and modern China to the world, developing a framework for the conservation and inheritance of Beijing's historic and cultural heritage with conservation as the top priority and inheritance as the foremost concern, and telling the stories of China, the Communist Party of China, and Beijing in a comprehensive and authentic manner to support the identity system of Chinese civilization.

Firmly bearing in mind the new development philosophy, properly managing the relationships between cultural heritage conservation and utilization, and between cultural heritage conservation and socio-economic development, coordinating conservation, inheritance and utilization, promoting creative transformation and innovative development of fine traditional Chinese culture, building a strong cultural foundation for Chinese modernization, and contributing Beijing's strength to carrying forward Chinese cultural tradition and building a country with a strong socialist culture.

II   Planning References

1. *Hold High the Great Banner of Socialism with Chinese Characteristics Striving in Unity to Build a Modern Socialist Country in All Respects—*

Report to the 20th National Congress of the Communist Party of China (2022)

2. *Guidelines on Strengthening the Preservation and Inheritance of Historical and Cultural Heritage in Urban and Rural Construction by the General Office of the CPC Central Committee and the General Office of the State Council* (2021)

3. *National Territorial Spatial Planning Outline (2021-2035)* (2022)

4. *Several Opinions of the General Office of the CPC Central Committee and the General Office of the State Council on Strengthening Reform on Cultural Heritage Conservation and Utilization* (2018)

5. *Law of the People's Republic of China on Protection of Cultural Relics* (Revised in 2024)

6. *Intangible Cultural Heritage Law of the People's Republic of China* (2011)

7. *Regulations on the Conservation of Historic Cities, Towns and Villages* (Revised in 2017)

8. *Regulations on the Conservation of Beijing Historic City* (2021)

9. *Beijing Urban Master Plan (2016-2035)* (2017)

10. *Regulatory Plan for the Core Area of the Capital City of Beijing (Block Level) (2018-2035)* (2020)

Ⅲ Temporal and Spatial Framework

The scope of this plan aligns with the administrative boundaries of Beijing, covering a total area of 16,410 square kilometers.

The planning period extends from 2023 to 2035, with the short-term planning set until 2027 and the long-term planning until 2035.

Ⅳ Planning Principles

Adhere to overall planning and systematic advancement. Upholding the leadership of the Party and the coordination of the government, strengthening top-level design, establishing a scientifically categorized, robustly protected and effectively managed framework for the conservation and inheritance of Beijing's historic and cultural heritage in urban and rural areas. Ensuring comprehensive integration of all time periods, full coverage of space, inclusion of all necessary elements, conservation throughout the entire process and full societal participation. Coordinating planning, implementation and management processes, and fostering the integration of conservation and inheritance of historic and cultural heritage and urban and rural development.

Adhere to a value-oriented approach and ensure the preservation of what should be preserved. From the perspectives of heritage and civilization, identifying core values of Beijing's historic and cultural heritage to support the identity system of Chinese civilization, providing effective support for accelerating the development of the Chinese discourse and narrative system. In line with the principles of authenticity and integrity, adapting to the unique conditions for the conservation of historic and cultural heritage, projecting and highlighting symbols of Chinese culture and images of the Chinese nation in urban and rural development, and promoting and developing the fine traditional Chinese culture, revolutionary culture and advanced socialist culture.

Adhere to reasonable use, inheritance and development. Upholding a people-centered approach, integrating heritage conservation and inheritance into economic and social development, ecological civilization building and modern life, and fulling leveraging the cultural, social and educational value of historic and cultural heritage. Encouraging proactive and reasonable use, fostering creative transformation and innovative development, focusing on improving people's livelihoods, and meeting the growing demands of the people for a better life.

Adhere to multi-party participation for synergy. Encouraging and guiding non-governmental sectors to actively engage in conservation and inheritance efforts, and fully leveraging the role of the market and stimulating the initiative and enthusiasm of the people to establish systems and mechanisms and foster a social environment conducive to the conservation and inheritance of historic and cultural heritage.

V Planning Objectives

By 2027, Beijing's historical and cultural heritage properties in urban and rural areas will be thoroughly identified, inventories of protected heritage properties continuously developed, a conservation and heritance framework established, a number of replicable and scalable experiences for conservation and sustainable use fostered, and conservation and inheritance efforts integrated into urban and rural development, contributing to the high-quality development of the capital.

By 2030, surveys of various historic and cultural heritage resources will be completed and inventories of protected heritage properties established, and the preliminary framework for the conservation and inheritance of Beijing's historic and cultural heritage in urban and rural areas will be developed. The conservation and inheritance of historic and cultural heritage will be fully integrated into the urban and rural development framework. Beijing's conservation priorities will be more prominent in the identity system of Chinese civilization. The interpretation and presentation of core values of Beijing's historic and cultural heritage will be strengthened. The cultural visibility, transmissibility and approachability of Beijing as the capital of the great nation will be further enhanced.

By 2035, a conservation and inheritance framework for Beijing's historic and cultural heritage will be established in all respects, and conservation and inheritance efforts will be deeply embedded in the broader context of urban-rural development and socio-economic growth. Beijing will be built into a world-renowned cultural city that showcases cultural confidence, diversity and inclusiveness, highlighting the continuity, innovation, unity, inclusiveness and peaceful nature of Chinese civilization.

Ⅵ Main Tasks

First, identify value attributes of Beijing's historic and cultural heritage in urban and rural areas to support the identity system of Chinese civilization and develop a value-based overall framework for conservation and inheritance.

Second, identify key aspects of conservation and inheritance of historic and cultural heritage in urban and rural areas, adhere to the principle of prioritizing protection and giving precedence to inheritance, focus on value interpretation, and establish a comprehensive conservation framework in addition to developing innovative inheritance approaches.

Third, strengthen protection and management and improve the implementation mechanism for conservation. This involves aligning with major national strategies and related national economic and social development plans and territorial spatial plans and setting out recommended conservation and inheritance projects.

Chapter 2

Developing a Value Framework to Guide Conservation and Inheritance

I   Overview of the Cultural and Historical Context

Beijing is located near the 40° N latitude line, at the junction of the second and third terraces of China's topography, to the southeast of the 400mm precipitation line and the transitional zone between agricultural and pastoral areas. The Beijing Plain, located at the northern edge of the North China Plain, is where three major geographical units converge, including the Inner Mongolian Plateau, the North China Plain, and the Northeast Region.

Beijing lies within the temperate monsoon continental climate zone, with diverse and complex topography and climate conditions. It is bordered by the Taihang Mountains to the west and the Yanshan Mountains to the north. The city is nurtured by five major river systems: the Yongding River, the Chaobai River, the North Canal, the Ju River, and the Juma River, which together provide abundant ecological resources.

Beijing preserves the earliest traces of human activity from the Paleolithic period. The Peking Man Site at Zhoukoudian and its surrounding areas have yielded remains of the Peking Man from around 700,000 years ago, the New Cave Man from 100,000 to 200,000 years ago, the Tianyuan Cave Man from 40,000 years ago, and the Upper Cave Man from 10,000 to 30,000 years ago. These findings span three stages of ancient humans: Homo erectus, early Homo sapiens, and late Homo sapiens. The Donghulin Site in Mentougou, dating from the early Neolithic period (11,000-9,000 years ago), is one of the origins of millet cultivation. Millet and broomcorn starch grains as well as tools unearthed from the Shangzhai Ruins in Pinggu (dating from 7,500-6,000 years ago) provide solid evidence that dryland farming had already emerged in Beijing during the middle and late Neolithic periods.

Beijing is located at the crossroads of various livelihood zones and diverse ethnic groups. Over the millennia and centuries, it has evolved from a vassal capital on the periphery of the Huaxia civilization sphere during the Western Zhou and Eastern Zhou dynasties, to a strategic frontier stronghold in the Qin and Han dynasties, and later, to a capital since the Liao and Jin dynasties. The Liulihe Ruins is the earliest known urban settlement in Beijing. A series of archaeological discoveries at the site have substantiated the historical records in the *Shiji* (Records of the Grand Historian), which mention the enfeoffment of Duke Zhao in the northern Yan during the early Western Zhou dynasty. The inscription of "Taibao Yongyan" on a bronze ware at the Liulihe Ruins clearly identifies Duke Zhao as the builder of the capital of Yan during the Western Zhou dynasty. Since then, Beijing's urban history has continued uninterrupted for over 3,000 years. In 1153, when the Jin dynasty moved its capital to Yanjing and renamed it Zhongdu, Beijing became the capital of a northern ethnic dynasty that controlled the core regions of the Central Plains, marking the beginning of over 870 years of the city's history as a national capital. By the Yuan dynasty, Beijing had ascended to the political and cultural center of the unified country, a role that was further solidified during the Ming and Qing dynasties.

Beijing is the birthplace of the Self-Strengthening Movement and the 1898 Hundred Days' Reform in the Qing Dynasty, as well as the national center of the New Culture Movement and the May Fourth Movement in the 1910s and 1920s. It was also a key forefront for the early spread of Marxism in China and one of the primary birthplaces of the Communist Party of China (CPC). In China's modernization process, Beijing played an irreplaceable and pivotal role.

After the founding of the People's Republic of China, Beijing underwent a remarkable transformation from the capital of a traditional society to that of a modern republic. Under the leadership of the CPC, the city embarked on capital construction projects represented by the "Ten Great Buildings" and the renovation of Tiananmen Square. Following the Reform and Opening-up policy since 1978, Beijing initiated a reform in its economic structure. In the 1990s, the city further strengthened its role as the national political and cultural center. The 2008 Beijing Olympic Games significantly enhanced China's global visibility. Since the 18th CPC National Congress, the CPC Central Committee has defined Beijing's strategic position as the "Four Centers": the national political center, cultural center, center for international exchanges, and center for technological innovation. Marked by the development of Beijing Municipal Administrative Centre in Tongzhou and the hosting of the 2022 Winter Olympics, Beijing is evolving into a great capital for the rejuvenation of the Chinese nation and a world-class city of harmony and livability.

II Developing a Value Framework from Multiple Perspectives

Chinese civilization has a time-honored history. The Master Plan focuses on China's million-year history of human activity, 10,000-year history of culture, over 5,000-year history of civilization, over 180-year contemporary history, over 100-year history of the CPC, over 70-year history of the People's Republic of China, and over 40-year history of reform and opening-up.

Guided by the unified national value framework and considered from multiple perspectives, including political, economic, social, technological, cultural, geographic and ecological, the Master Plan systematically examines Beijing's significant position in the historical development of Chinese civilization to establish a core value framework for Beijing's historic and cultural heritage, identify priority protected sites and monuments to reconstruct a value-based conservation mechanism, and highlight Beijing's distinctive historical and cultural features through comparative study with other ancient capitals worldwide and within China.

Ⅲ Core Value Framework of Beijing's Historic and Cultural Heritage

1. Summary of Overall Value

Beijing, with its time-honored history and enduring cultural lineage, stands as a powerful testament to the continuity, innovation, unity, inclusiveness and peaceful nature of Chinese civilization.

Beijing's unique geographical layout and geological features are remarkable evidence of early human-environment relationships in East Asia. It is the culmination of ancient Chinese urban planning, a thousand-year-old capital that bears witness to the continuous evolution of Chinese civilization. Beijing is also a monumental witness to the transformation of China's political system in the modern era, serving as the national capital guiding China's path to modernization. It has witnessed the collision, interaction and fusion of different civilizations, peoples and cultures across the Eurasian continent, making it a great cosmopolitan capital of openness, diversity and inclusiveness.

When compared with other contemporary capitals worldwide, Beijing exemplifies the ritual-based urban planning philosophy of capital cities, representing the highest achievements of oriental civilization.

2. Value Themes and Attributes

(1) Value Ⅰ: An exceptional testimony to early human-environment relationships in East Asia

Beijing, located at the junction of the second and third terraces of China's northern topography, preserves a great wealth of geological landscapes, remains of human and agricultural origins, and diverse wild flora and fauna. It stands as an exceptional example of early human-environment relationships in East Asia.

Value Attribute I-1: The Peking Man remains represented by the Peking Man Site at Zhoukoudian are of milestone significance in the history of human evolution.

Value Attribute I-2: The sites and remains represented by the Donghulin Site serve as an outstanding testament to the origins of millet farming in Northeast Asia.

Value Attribute I-3: The geological remains in Fangshan and Yanqing areas of Beijing are the most typical representations of the karst landform in northern China and the Yanshan Mountains movement. The abundant wild flora and fauna in the western and northern mountainous areas exhibit the biodiversity of North China.

(2) Value II : A millennium-old capital witnessing the continuity of Chinese civilization

Since the 12th century, Beijing has been the national capital of the Jin, Yuan, Ming and Qing dynasties in the later period of ancient Chinese society. It is the final form that presents the evolution of capital planning ideals and ritual-based spatial order in ancient China, embodying the greatest accomplishments of urban civilization in East Asia during this period.

Beijing relies on the Yanshan Mountains, which nestle between the Greater Khingan Range and the Taihang Mountains, holding a significant geographical and cultural position in China. Its strategic location allows for communication across various regions, linking agricultural, pastoral, fishing and hunting zones, and creating a strategic pattern of "keeping control over the Central Plains to the south and connecting the northern frontier to the north". It serves as the geographical foundation for a unified multi-ethnic country's capital and its human-environment relationships. The Great Wall defense system and the urban layout, shaped by the spatial influences of different historical periods, bear witness to the collision and integration of diverse peoples and cultures.

Value Attribute II-1: Beijing's geographical location holds a unique strategic value in the process of China's national unification in ancient times, serving as the critical geographical foundation for the capital of a unified multi-ethnic country.

Value Attribute II-2: The city's urban construction since the Western Zhou dynasty, along with the capital city development by the Jin, Yuan, Ming and Qing dynasties in the later period of ancient Chinese society, is a great testament to the development of a unified multi-ethnic country in Chinese civilization.

Value Attribute II-3: The spatial planning layout of Beijing's Old City embodies the classical paradigm of ancient Chinese capital city planning, based on the ritual system.

Value Attribute II-4: The layout of Beijing's ancient ritualistic architectural complexes demonstrates the systematic and characteristic role of Chinese civilization in establishing and maintaining the traditional societal order.

Value Attribute II-5: Beijing's imperial palaces, temples, gardens and mausoleums represent the finest examples of royal architecture in ancient China, exhibiting a complete series of masterpieces that reflect traditional Chinese court culture.

Value Attribute II-6: Beijing's Great Wall defense system is a unique testament to the process of collision and integration between agrarian culture and nomadic, fishing and hunting cultures in Chinese history.

Value Attribute II-7: Beijing's canal system bears witness to the development of the *caoyun* (tribute grain transport) system and a series of innovations in hydraulic engineering technology, ensuring the resource supply necessary for Beijing as the capital of a unified multi-ethnic country in ancient times.

(3) Value Ⅲ: A leading capital of excellence in Chinese modernization

Beijing has undergone profound changes in modern history, witnessing unprecedented transformations that have reshaped China. It was a central forefront for the early spread of Marxism in China and one of the primary birthplaces of the Communist Party of China. Throughout China's transition from tradition to modernity in the 20th century, Beijing has witnessed the transformation of the political system, ideological and cultural shifts, social and economic changes, and the evolution of urban functions and development models. From a dynastic imperial capital to the capital of the Republic, Beijing has become the leading capital of excellence in China's modernization process.

Value Attribute Ⅲ-1: Beijing has witnessed regime changes, ideological clashes, revolutionary processes, and urban transformations, witnessing the shift from a dynastic imperial capital to the modern capital of a republic.

Value Attribute Ⅲ-2: After the founding of the People's Republic of China in 1949, the capital construction in Beijing under the leadership of the Communist Party of China became an outstanding example of the Chinese people's advancement in socialist revolution and construction.

Value Attribute Ⅲ-3: After the Reform and Opening-up in 1978, Beijing took new strides in development, continuously strengthening its functions as "Four Centers" and highlighting its leading role in advancing the great rejuvenation of the Chinese nation through a Chinese path to modernization.

(4) Value IV: A great capital of diversity, openness and inclusiveness

Over the past thousand years, Beijing has witnessed the rule of various ethnic groups, including the Khitan, Jurchen, Mongols, Han and Manchu. These different groups brought their distinct customs, cultures, religious beliefs and scientific ideas, which collided, interacted and integrated. In this way, Beijing has developed a multifaceted urban culture, embodying the open and inclusive nature of the capital of a great nation.

Value Attribute IV-1: Beijing was where China's various ethnic groups interacted and integrated their cultures, exhibiting the open and inclusive nature of the capital of a great nation.

Value Attribute IV-2: The remains of different religious beliefs in Beijing, such as Buddhism, Taoism, Islam, Christianity and Catholicism, along with folk traditions, reflect the city's inclusive character as the capital of a great nation.

Value Attribute IV-3: Beijing's traditional *hutong* alleys, courtyard compounds, guildhalls, shops, teahouses, theatres, villages, and agricultural and handicraft sites or workshops, as well as their historical surrounding settings and vegetation, exhibit the distinctive cultural tradition of Beijing.

Value Attribute IV-4: The various public buildings and urban facilities in Beijing, shaped by exchanges of Eastern and Western ideas, cultures, and scientific and technological advancements, bear witness to the transformation of production patterns, lifestyles and urban development models.

Value Attribute IV-5: The remains of China's top educational institutions and cultural organizations since the Yuan dynasty, along with numerous residences of notable figures, testify to the evolution of China's educational system and stand as a culmination of diverse thoughts and cultures.

Chapter 3

Strengthening Systematic Conservation to Secure Heritage Resources

I   Strengthening the Management of Heritage Listings

1. Improving the framework of conservation objects

The conservation objects to be protected under the framework of the conservation and inheritance of Beijing's historic and cultural heritage in urban and rural areas primarily include both urban and rural, natural and cultural, tangible and intangible heritage that possess significant conservation value and carry the cultural connotations of different historical periods, as well as their contextual environments. These protected heritage properties range in the following categories: a. World Cultural Heritage Sites; b. cultural relics; c. historic buildings [including excellent buildings of early modern and modern times, industrial heritage properties, listed courtyards, and old (former) residences of famous people] and revolutionary sites; d. historic conservation areas, areas of special quality, and underground archaeological remains; e. historic towns and villages and traditional villages; f. historic rivers and lakes and aquatic cultural heritage properties; g. mountain-and-river-based patterns and city-site remains; h. traditional *hutongs*, historic streets and alleys, and traditional place names; i. scenic areas, historic gardens, and ancient and valuable trees; j. intangible cultural heritage; and k. other protected heritage specified by laws and regulations.

By identifying areas that may lack sufficient support for the core value framework, evaluating gaps in the value carriers, and leveraging results from surveys of cultural heritage and historical and cultural resources, potential heritage properties will be further identified and added to heritage listings when conditions are met.

2. Identifying conservation priorities

A value-oriented approach should be employed to sustain ongoing systematic research on value carriers, catalog historical and cultural resources, and identify priorities for the conservation of Beijing's historic and cultural heritage based on the value framework.

Value Framework of Beijing's Historic and
Cultural Heritage and Its Conservation Priorities   Table 1

Value Theme	Value Attribute	Conservation Priority
An exceptional testimony to early human-environment relationships in East Asia	I-1 Human origins	Significant sites of human and agricultural origins
	I-2 Origins of agriculture	
	I-3 Geological features and biodiversity	Important geological heritage landscapes
A millennium-old capital witnessing the continuity of Chinese civilization	II-1 Geographical pattern, providing the geographical foundation for the capital of a unified multiethnic country in Chinese civilization	Geographical landscape of the national terrains
	II-2 The history of Beijing as a city and a capital, bearing extraordinary testimonies to the development of a unified multiethnic country in Chinese civilization	Important city-site remains in Beijing as well as capital sites, including Zhongdu of the Jin dynasty, Dadu of the Yuan dynasty, and the historic center of Beijing in the Ming and Qing dynasties
	II-3 The classical paradigm of capital city planning in ancient China	Distinctive features of the urban spatial layout in China's late imperial period, primarily including the four-tiered layout framework of the palace city, the imperial city, the inner city and the outer city symmetrically arranged along the central axis, as well as the grid-like road network, from the Ming and Qing dynasty
	II-4 Ancient ritualistic architectural complexes, embodying the cultural essence of China's ritual system	Temples and altars related to ancient state and imperial ritual activities

Continued

Value Theme	Value Attribute	Conservation Priority
A millennium-old capital witnessing the continuity of Chinese civilization	II-5 The most representative examples of ancient Chinese imperial architecture	Imperial buildings, including palaces, gardens, mausoleums and temples
	II-6 The Great Wall defense system	Sites and remains related to the Great Wall defense system in Beijing, including border walls, beacon towers, passes, fortresses, inscriptions, and other associated facilities
	II-7 The canal system	Sites and remains related to the canal system in Beijing, including rivers, lakes, waterworks, bridges and storage facilities
A leading capital of excellence in Chinese modernization	III-1 Regime changes, ideological clashes, revolutionary processes in modern times	Locations of government offices and places associated with revolutionary activities from 1840 to 1949
	III-2 Construction of the capital after the founding of the People's Republic of China in 1949	Representative buildings and major projects reflecting the construction of the capital from 1949 to 1978
	III-3 Great rejuvenation after the Reform and Opening-up in 1978	Representative buildings and major projects reflecting the development of the capital and its functions as "Four Centers"
A great capital of diversity, openness and inclusiveness	IV-1 Cultural interaction and integration among various ethnic groups	Architectural remains of ethnic minorities for living and production purposes
	IV-2 Diversity of religious beliefs	Representative religious places and their sites and remains, including Buddhism, Taoism, Islam, Christianity, Catholicism, and folk beliefs
	IV-3 Local cultural traditions	Representative *hutongs*, courtyards, guild halls, shops, markets, teahouses, theatres, villages, and farming and handicraft places as well as their historical surrounding settings and vegetation

Continued

Value Theme	Value Attribute	Conservation Priority
A great capital of diversity, openness and inclusiveness	Ⅳ-4 Intellectual, cultural and technological exchanges between East and West	Representative modern buildings (building complexes) for administrative, diplomatic, exhibition, financial, medical care, educational, industrial and transportation functions, as well as city parks
	Ⅳ-5 Diversity of educational ideas and cultures	Ancient and modern educational facilities and various old (former) residences of famous people

3. Dynamically managing heritage listings

Hierarchical management should be strengthened. The management of heritage properties at the national and municipal levels should be strengthened, focusing on their historical and cultural values. Designations of World Cultural Heritage Sites, Heritage Sites Protected at National Level, Traditional Villages of China, Historic Conservation Areas of China, National Industrial Heritage, National Representative Intangible Cultural Heritage, China Time-Honored Brands, and Important Agricultural Heritage of China should be continuously advanced to ensure priority conservation of important value carriers and their inheritance and utilization.

The heritage listing management system should be implemented. In accordance with the *Provisional Regulations for the Identification and Registration of Protected Objects of Beijing Historic City*, efforts should be made to strengthen prior protection and promptly include heritage properties that possess conservation value and meet designation criteria into heritage listings. If protected heritage properties are damaged or lost due to force majeure, or undergo changes in terms of protective levels or types, adjustments should be made to corresponding heritage listings.

II Promoting Comprehensive and Systematic Conservation

1. Overall Conservation Requirements

(1) Strengthening conservation of cultural heritage from different periods

Conservation of heritage properties from ancient times should be strengthened. Historic sites that exhibit the history of urban constructions and multi-ethnic exchanges in ancient times and cultural landscapes that bear historical and cultural significance will be protected. The focus should be placed on strengthening the conservation of tangible evidence related to the origins of humanity, agriculture and other aspects of civilization.

Value of protected heritage properties from modern times should be explored. Efforts will be made to fully explore and protect heritage properties that reflect historical processes or outstanding accomplishments in modern wars and conflicts, revolutionary movements, political system reforms, industrial and commercial development, changes in lifestyles, the spread of new ideas and cultures, scientific and technological progress, and urban and architectural development. The focus should be placed on the conservation of memorial sites related to modern wars and conflicts and political system reforms, such as the Self-Strengthening Movement, the 1898 Hundred Days' Reform, the Boxer War, the 1911 Revolution, the May Fourth Movement, and the Lugou Bridge (Marco Polo Bridge) Incident.

Further efforts should be made to protect heritage properties that reflect the CPC's journey of uniting and leading the Chinese people in their relentless struggle, such as sites and monuments that commemorate the CPC's early revolutionary activities, the establishment of anti-Japanese revolutionary bases, and the founding of the People's Republic of China. Key heritage properties to be protected include the Red Building of Peking University and its surrounding area, the site of the February 7th Strike in Changxindian, the anti-Japanese bases in Beijing's suburbs, and the

Xiangshan Revolutionary Memorial Site in Beijing. Relevant memorial sites should be established to ensure the revolutionary legacy is passed on.

Efforts should be made to strengthen the conservation of heritage properties in Beijing that showcase significant historical events, such as the founding of the People's Republic of China, its early construction, and the Reform and Opening-up, and tell the story of Beijing in the context of the People's Republic of China. The value features of heritage properties that reflect the establishment of China's political system, the superiority of socialist system, and achievements in urban construction, technological innovation, social development, and international exchanges should be further explored.

(2) Strengthening systematic conservation of regional cultural heritage

Efforts should be made to promote the systematic conservation of regional historic and cultural heritage, with focus on the construction of the National Culture Parks of the Great Wall and the Grand Canal and the conservation of linear heritage sites, such as the Yongding River, the Beijing-Zhangjiakou Railway, and the Ancient Road in Western Beijing.

Comprehensive conservation of areas rich in historical and cultural resources should be strengthened to protect various heritage properties and their contextual settings within these areas. Associations between these elements should be fully explored and cultural nodes and cultural routes should be developed to coordinate and interconnect heritage properties.

(3) Strengthening the comprehensive conservation of fabrics of heritage properties and their natural and cultural settings

Comprehensive conservation of the Taihang Mountains and the Yanshan Mountains as well as major rivers and valleys, including the Yongding River, Chaobai River, Juma River, North Canal, and Ju River, should be carried out. Ecological management in mountainous areas should be strengthened and ecological conservation and heritage preservation should be coordinated to prevent natural disasters, such as water loss and soil erosion, ground subsidence, earthquakes and mudslides, from damaging heritage properties. Arbitrary alterations or encroachments upon rivers and lakes shall not be allowed, while enhancing the waterfront environment. Value of the heritage's contextual settings should be highlighted. Efforts should be intensified to protect and manage mountains, rivers, farmlands and forests that have been included in conservation areas or construction control zones. Features of natural environments and regional cultures where protected heritage properties rest on should be maintained, and necessary visual corridors should be established to restrict development projects.

Tangible cultural heritage properties such as natural environments, spatial layouts and textures, traditional buildings, and historical setting elements of cities, towns and villages, and intangible cultural heritage items such as folk arts and local customs and traditions as well as their cultural ecosystems, should be protected in an integrated manner. The integrity of historical features, the continuity of social life, and the diversity of urban and rural functions should be maintained to preserve historical and cultural contexts, ensuring that the historical traces of different eras and memories of daily life are adequately protected.

2. Categorized conservation priorities

The goal is to highlight the core value of protected heritage properties, emphasizing comprehensive and systematic conservation based on the "value-carrier-environment" framework. Differentiated conservation measures should be formulated in line with characteristics of each type of protected heritage properties.

(1) **World Cultural Heritage Sites.** Strictly adhere to the requirements of the *Convention Concerning the Protection of the World Cultural and*

Natural Heritage, safeguarding the authenticity and integrity of World Cultural Heritage Sites and further strengthening the interpretation and presentation of outstanding universal value. Implement specialized conservation management plans and enhance protection and management. Select potential World Heritage nominations and use the nomination process as a lever to promote the conservation of historic and cultural heritage and drive regional conservation and renewal.

(2) **Cultural relics.** Strictly implement relevant laws and regulations regarding the conservation of cultural heritage. Adhere to the principles of "prioritizing preservation, acting swiftly to rescue, utilizing rationally, and strengthening management". Strengthen preventive conservation and regular monitoring and management of cultural relics, with a focus on addressing issues such as improper use of cultural relics and disharmonious surroundings. Actively promote the vacation of improperly used cultural relics and their rational utilization, while rectifying non-compliant construction activities within construction control zones for officially protected sites. Implement the requirements of "protection first, strengthening management, exploring values, utilizing efficiently, and bringing cultural relics to life". Encourage the integration of heritage conservation with value interpretation and presentation, reuse for public benefit, and transmission of intangible cultural heritage.

(3) **Historic buildings.** Strengthen the preservation and repair of historic buildings, based on relevant conservation standards and under the premise of not altering the core value attributes. Develop technical standards and funding subsidy policies for repair works to promote adaptive reuse.

(4) **Revolutionary sites.** Highlight major historical events and figures, maintain the safety of fabrics of revolutionary sites and their distinctive historical features. Organize commemorative activities and collecting, sorting out and exhibiting historical documents and sites associated with

heroic figures. Encourage the integration of revolutionary sites with other protected heritage properties, expanding display routes and contents to develop thematic trails.

(5) **Historic conservation areas.** Promote the development of conservation plans for historic conservation areas. Place a priority on the conservation of historic layouts, street and alley fabrics, spatial scale, and landscape environment of historic conservation areas. A categorized approach should be applied to the conservation of buildings within these areas, and disharmonious buildings and landscapes should be gradually rehabilitated to continue historic features. Maintain the continuity of daily life, carry forward cultural traditions and customs, preserve the way of life and community environment within historic conservation areas. While meeting conservation requirements, enhancing infrastructure and public service facilities and improving living conditions in historic conservation areas.

(6) **Areas of special quality.** Sustain main functions and distinctive features of areas of special quality, preserve historic and cultural heritage and their environmental elements that reflect core values, protect existing buildings that represent specific stages of urban development, and enhance their cultural display functions. For areas of special quality focused on residential functions, protect their street and alley fabrics and traditional features, preserve the elements of historical environments that carry the memories of residents and emotions of citizen groups, optimize infrastructure, and improve the living environment for residents. For areas of special quality focused on economic and industrial functions, retain their core production space layouts, advance the rehabilitation of surrounding environments, appropriately add cultural display functions, and adopt differentiated measures such as continuity, upgrading or replacement of industries based on the current industrial situation. For areas of special quality focused on scientific, educational and cultural functions, protect their spatial layouts, traditional features, historic environments,

gardens and landscapes, while carrying forward the cultural spirit they embody.

(7) Underground archaeological remains. Strictly adhere to the requirements outlined in the *Administrative Measures of Beijing Municipality for the Conservation of Underground Archaeological Remains*. Explore a graded management approach for underground archaeological remains. Development projects to be carried out in key areas should undergo careful evaluation before project initiation. For sites that may contain valuable ancient ruins or tombs, conservation plans should be developed in advance, ensuring comprehensive preservation of underground cultural objects. For general areas, key monitoring areas, and areas outside known underground archaeological remains, the principle of "archaeology taking priority over development" should be followed, with development schemes established based on the results of archaeological activities.

(8) Historic towns and villages and traditional villages. Push forward the preparation of conservation plans for historic towns and villages, conservation and development plans for traditional villages, and conservation and utilization plans for contiguous traditional villages. Protect the overall spatial form of town and village sites, including topography, landscape environment, and historical layouts. Protect historic remains, elements of historic environments, and distinctive vernacular houses that highlight value features. Protect and continue agricultural elements that sustain these towns and villages, such as farmlands, woodlands and ponds. Protect traditional ways of production and living, folk customs and intangible cultural heritage. Adopt a use-driven conservation approach to revitalize traditional buildings, promote the renovation of traditional houses to improve their livability. Enhance infrastructure and address gaps in public service facilities to improve the quality of the living environment. Explore and carry forward traditional cultural practices such as folk customs and develop related industries like cultural tourism. Strengthen the construction of the safety and security system, improving fire protection, disaster prevention and other safety facilities.

(9) Historic rivers and lakes and aquatic cultural heritage. Protect the overall course of historic rivers and lakes, maintain the original shape of the waterways and traditional embankments wherever possible, and gradually restore historic river channels that bear significant value for urban development. Manage the use of aquatic cultural heritage properties such as bridges, dams and sluices in a reasonable way. Improve facilities like water conservancy museums and exhibition halls, enhancing the live exhibition of aquatic culture.

(10) City-site remains. Ensure the comprehensive conservation of ancient city walls, gates and moats, preserving the historical layout of the Old City. Protect road fabrics and key axes as well as important node remains within the Old City.

(11) Traditional *hutongs*, historic streets and alleys, and traditional place names. Strengthen the conservation of traditional *hutongs* and historic streets and alleys, exercising priority protection of those with preserved features and dimensions, as roads that should never be widened. Maintain the fabrics of traditional streets and alleys and their spatial scale. Strengthen the inheritance and management of place name heritage properties, retaining or prioritizing the use of traditional place names during block conservation and renewal.

(12) Scenic Areas. Strictly implement the *Regulations on Scenic Areas*, protect the natural environment and cultural landscape of scenic areas, promote wholesome sightseeing, cultural and recreational activities, and disseminate historical, cultural and scientific knowledge.

(13) Historic gardens and ancient and valuable trees. Preserve the spatial layout and historical appearance of historic gardens to the greatest possible extent, and protect historic environment elements such as rocks, water features, and ancient and valuable trees. Strictly implement the *Regulations of Beijing Municipality for the Conservation of Ancient and Valuable Trees* and its implementation guidelines, preserving ancient and valuable trees and their habitats and strengthening the publicity of the historical, scientific and cultural values of these trees.

(14) Intangible cultural heritage. Strictly implement relevant laws and regulations on the safeguarding of intangible cultural heritage. Following the principle of "keeping people, objects and life visible", safeguarding and transmiting intangible cultural heritage and the cultural and natural environments that sustain them. Strengthen the development and training of ICH bearers. Integrate intangible cultural heritage into tangible cultural heritage, and reasonably planning experience spaces for the publicity, exhibition, transmission of intangible cultural heritage.

(15) Time-honored brands. Strengthen the protection of time-honored brand heritage. Enterprises of time-honored brands that possess cultural properties with significant value and their original business locations that meet the required conditions should be designated as immovable cultural heritage by law. Preservation of the original site and the appearance of time-honored brand businesses should be strengthened. In case of relocation, comments of the competent authorities shall be sought. If relocation is necessary, compensation should be provided according to the principle of "demolish one, compensate one" or based on the assessed value of the original property, ensuring it is no less than the valuation of the original ownership entity.

(16) Agricultural heritage system. Identify more important agricultural heritage systems, protect and preserve their traditional modes of production, lifestyles, and associated cultural customs, and establish core conservation areas. Integrate ecological environment protection, village development and agricultural heritage conservation efforts.

Ⅲ Coordinating Planning and Control

Strictly implement the conservation and control requirements for various types of historic and cultural heritage, and consolidating conservation areas and construction control zones. Strengthen the alignment of territorial spatial plans with conservation and inheritance requirements, properly handle the relationship between conservation areas and construction control zones of historic and cultural heritage properties and the "Three Areas and Three Lines" as well as conservation boundaries, incorporating them into the unified map of territorial spatial planning to enhance planning and control.

Within urban development boundaries, prevent large-scale demolition and construction that could damage various types of protected heritage properties and their settings. Clearly define regulatory requirements and guiding measures to avoid concentrated construction that negatively impacts the fabrics of historic and cultural heritage properties and their settings. Coordinate cultural heritage conservation and urban and rural development.

Encourage the coordinated conservation and utilization of permanent basic farmlands and various types of protected heritage properties. For permanent basic farmlands within large-scale archaeological sites and underground archaeological remains, agricultural activities may proceed in good order without damaging the site if there are no ongoing archaeological plans. If there are archaeological plans, temporary land use approval for archaeological activities should be obtained. If agricultural land needs to be converted into construction land for purposes such as conservation and presentation, the required construction land approval procedures must be followed. Strengthen the conservation of cultural heritage during planned agricultural production activities, such as the development of high-standard farmlands. Planned agricultural activities should, as much as possible, avoid areas with immoveable cultural heritage.

Strengthen the conservation and restoration of regional ecological environments that align with the conservation of historic and cultural heritage within the ecological conservation redlines. Develop guiding measures for ecological restoration works. Without damaging ecological functions, allow archaeological activities, conservation efforts for protected heritage properties, as well as moderate tourism activities and construction of necessary public facilities, within ecological conservation redlines but outside core conservation areas of the Protected Areas , provided that they shall be approved by law. This will aim to promote the rational use of cultural and natural heritage properties.

IV Establishing a Spatial Framework for Conservation and Inheritance

1. Identifying features of the overall spatial framework for historic and cultural heritage

The Master Plan aims to establish a spatial framework for the conservation and inheritance of Beijing's historic and cultural heritage in urban and rural areas that features "Two Mountains and One Bay, Two Areas and Two Axes, Three Belts and Multiple Zones", based on Beijing's historical context and evolution features of its urban spatial layout, guided by historical and cultural value, with patterns of physical geography and characteristics of diverse regional cultures as the foundation, key cultural corridors and routes as linkages, and distribution features of historic and cultural heritage as supportive elements.

"Two Mountains and One Bay" refers to the Taihang Mountains, the Yanshan Mountains, and the Beijing Bay, which form the natural foundation of Beijing's historical transformations and cultural developments. They represent the origins of human activities, the root of history, and the basis for the establishment of the capital in Beijing.

"Two Areas and Two Axes" refers to two areas of the Old City and the Three Hills and Five Gardens, and two axes being the Central Axis and its extensions, and Chang'an Avenue and its extensions. Among these, the Old City and the Three Hills and Five Gardens are two key areas where traditional Chinese culture is concentrated and the development of Chinese modernization is most evident. The Central Axis and Chang'an Avenue span across time and space, bearing a direct testimony to the urban development of Beijing as a capital city from ancient to modern times.

"Three Belts and Multiple Zones" refers to the Great Wall Cultural Belt, the Grand Canal Cultural Belt and the Western Hills-Yongding River Cultural Belt; the Multiple Zones being the priority conservation and inheritance zones connected by these cultural belts. These areas represent a concentration of Beijing's historic and cultural heritage properties and their values.

2. Defining conservation requirements for different characteristic spaces

(1) Comprehensive conservation of the natural landscape of the Taihang and Yanshan Mountains

Properly protect the geographical heritage and landscape environment of the Tanghang and Yanshan Mountains. Conduct ecological environment assessments, explore ecological compensation mechanisms based on the preservation of history and culture, strengthen ecological restoration and environmental rehabilitation, safeguard biodiversity, and ensure the safety of ecological environment. Continue large-scale afforestation and greening efforts to restore historic and cultural landscapes. Exercise priority conservation of natural reserves, scenic areas, forest parks, geological parks, wetland parks, and other protected areas.

Properly preserve the spiritual and cultural lineages of the Taihang and Yanshan Mountains. Systematically comb through and deeply explore their rich cultural connotations, emphasizing the essence of traditional Chinese thought on the harmonious coexistence between man and nature.

(2) Comprehensive conservation of the Old City

The Old City of Beijing is an unparalleled masterpiece of Chinese urban planning. It stands as a crucial testament to the history of Chinese civilization, the course of modern Chinese history, the glorious journey of the Communist Party of China leading the Chinese people in their tireless struggle, the founding and development of the People's Republic of China, and the great journey of reform and opening-up and socialist modernization. Efforts should be made to strictly enforce the requirement that the Old City must not be further demolished, prioritize its conservation, and exercise its comprehensive conservation with a more proactive attitude and a more scientific approach.

a. Comprehensive conservation of the traditional spatial layout of the Old City. Protect the traditional Central Axis, the city layout of the historic center of Beijing in the Ming and Qing dynasties and the quadruple-walled city framework comprising the palace city, the imperial city, the inner city and the outer city; the imperial city of the Ming and Qing dynasties in an integrated manner, the structure of the chessboard road network and layout of streets, alleys and *hutongs*, the traditional architectural form of *hutongs* and *siheyuan* compounds, the gently expansive traditional urban spatial form, visual corridors and street view corridors, colors and shape features of traditional buildings, and ancient and valuable trees and big trees.

b. Protect ancient ceremonial and imperial building complexes in the Old City.

c. Protect representative heritage properties of the Old City that embody its diverse and inclusive local culture.

d. Protect representative modern heritage properties that showcase the transformation from an imperial capital of the dynastic era to a modern capital of the Republic.

e. Protect representative heritage properties that reflect the great achievements in the construction of the capital in the beginning years of the People's Republic of China, economic transformation during the Reform and Opening-up period, and development of the capital in the new era, under the leadership of the Communist Party of China.

(3) Comprehensive conservation of the Three Hills and Five Gardens

The Three Hills and Five Gardens is a composite area where traditional and emerging culture intertwine and interact. It is essential to systematically protect its historic and cultural heritage as well as natural landscapes, restore the natural features of its pastoral and mountainous sceneries, and create an comprehensive spatial structure that blends historic context with ecological environment, and traditional culture with modern civilization.

The "Three Hills and Five Gardens" collectively refers to cultural heritage from various historical periods, represented by the Qing dynasty imperial gardens located in the northwestern suburbs of Beijing. The "Three Hills" are Xiangshan (Fragrant Hills), Yuquanshan (Jade Spring Hill) and Wanshoushan (Longevity Hill), and the "Give Gardens" are the Jingyiyuan (Garden of Tranquility and Pleasure), Jingmingyuan (Garden of Tranquility and Brightness), Yuanmingyuan (Garden of Perfection and Brightness) and Changchunyuan (Garden of

Everlasting Spring) as well as the Summer Palace.

a. Comprehensive conservation of the landscape pattern in the Three Hills and Five Gardens.

b. Protect classical gardens, with imperial gardens at the core.

c. Protect the water system that connects the Old City and the Three Hills and Five Gardens, imperial roads and ancient villages along the routes, and the rice field landscape in western Beijing.

d. Protect significant heritage sites that reflect regime changes, ideological clashes and revolutionary processes in modern times.

e. Protect representative buildings that exhibit the development of education, science and culture since the founding of the People's Republic of China.

(4) Coordinated conservation and development of the "Two Axes"
The "Two Axes" (the Central Axis and its extensions, and Chang'an Avenue and its extensions) serve as the core elements guiding the organization of urban spaces and functions in Beijing. The traditional Central Axis, as a masterpiece exhibiting the ideal order of Chinese capitals, is a key area for state rituals and functions. Chang'an Avenue, as a testament to the historical progress of the great rejuvenation of the Chinese nation, stands as an important window showcasing the modernization of the capital.

a. Preserve and develop the "Two Axes" pattern formed by the Central Axis and Chang'an Avenue. Protect the spatial order of the Central Axis that embodies ceremonial harmony and symmetrical alignment, and the solemn, well-ordered and grand spatial form of Chang'an Avenue.

b. Protect important heritage properties along the Central Axis and its extensions that exhibit the ritual-based order of the nation.

c. Protect representative buildings along Chang'an Avenue and its extensions that embody the strategic status of the capital as the national political center.

d. Protect representative buildings along the "Two Axes" that reflect the construction and development of the capital from the founding of the People's Republic of China to the progress of the great rejuvenation of the Chinese nation through China's path to modernization.

(5) Coordinated Conservation and Development of the Great Wall, the Grand Canal and Western Hills-Yongding River Cultural Belts

The Great Wall and the Grand Canal are typical representations of the human-environment relationships in a large-scale unit of physical geography in ancient China, exhibiting an unparalleled unity of natural and cultural landscapes and standing as iconic symbols of the Chinese nation and significant emblems of Chinese civilization. The Yongding River, known as the "Mother River" of Beijing, is the root that nourishes Beijing's urban development and transformation. The three cultural belts centered around the Grand Canal, the Great Wall, and the Western Hills and Yongding River, are crucial spatial platforms and cultural ties that sustain the conservation and inheritance of Beijing's historic and cultural heritage.

The Great Wall Cultural Belt encompasses six districts: Mentougou, Changping, Yanqing, Huairou, Miyun, and Pinggu; the Grand Canal Cultural Belt spans seven districts: Changping, Haidian, Xicheng, Dongcheng, Chaoyang, Shunyi, and Tongzhou; the Western Hills-Yongding River Cultural Belt covers eight districts: Yanqing, Changping, Haidian, Shijingshan,

Fengtai, Mentougou, Fangshan, and Daxing.

a. Protect the Great Wall and its setting, including mountains, rivers and valleys, geological relics, defense structures, ancient villages, ancient roads, intangible cultural heritage and folk customs, cultural landscapes, and other natural and cultural heritage properties.

b. Protect the Grand Canal and its associated water conservancy facilities, ancient villages, intangible cultural heritage and folk customs, cultural landscapes, and other cultural heritage properties along its course.

c. Protect the geological relics, ancient roads, ancient sites, ancient villages, ancient buildings, intangible cultural heritage and folk customs, cultural landscapes, and other natural and cultural heritage properties along the Western Hills and Yongding River.

d. Protect representative buildings (complexes) and major projects that exhibit the revolutionary history in modern times, the construction of the capital after the founding of the People's Republic of China, and the great achievements during the Reform and Opening-up period, related to the Great Wall, the Grand Canal, and the Western Hills and Yongding River.

3. Coordinated conservation, inheritance and development of key areas

(1) Core values and conservation priorities of each area

Based on the spatial concentration and cultural association of various types of historic and cultural heritage properties, strengthen the spatial distribution features of historical and cultural value carriers. Outside the Old City and the Three Hills and Five Gardens, designate 14 distinctive, complementary, and concentrated key conservation areas, further refining the overall framework for the conservation of the city's historic and cultural heritage, guided by their historical and cultural values.

Guidelines for Key Conservation and Inheritance Areas   Table 2

District	Key Area	Core Value	Conservation Priority
	Olympic Common Domain-Dadu of the Yuan Dynasty	II-2 The history of Beijing as a city and a capital, bearing extraordinary testimonies to the development of a unified multi-ethnic country in Chinese civilization III-3 Great rejuvenation after the Reform and Opening-up	1. Protecting important physical carriers that exhibit construction practices of the ancient capital and their surrounding environment, such as the Yuan Dadu City Wall Ruins; 2. Protecting key venues that bear the functions of modern sports events and international exchanges, including the National Olympic Sports Center, the National Indoor Stadium, and the National Aquatics Center
Chaoyang District	Wangjing-Sanlitun-Beijing Central Business District	III-2 Construction of the capital after the founding of the People's Republic of China III-3 Great rejuvenation after the Reform and Opening-up	1. Protecting important sites that bear the history of modern industrial development, such as the historic building complex of the former Beijing Electron Tube Factory and the office buildings of the former Beijing Cable Factory; 2. Identifying and protecting core sites in the areas that play key roles in modern diplomacy and bear commercial functions, including the Sanlitun second embassy zone, the Liangmaqiao third embassy zone, the China World Trade Center
Fengtai District	Lianhuachi-Zhongdu of the Jin Dynasty	II-2 The history of Beijing as a city and a capital, bearing extraordinary testimonies to the development of a unified multi-ethnic country in Chinese civilization	Protecting important physical carriers that exhibit construction practices of the ancient capital and their surrounding environment, such as the Shuiguan Site from Zhongdu of the Jin Dynasty, the Site of Zhongdu of the Jin Dynasty, and Lianhuachi

Continued

District	Key Area	Core Value	Conservation Priority
Fengtai District	Lugou Bridge (Marco Polo Bridge) Changxindian	II-1 Geographical pattern, providing the geographical foundation for the capital of a unified multi-ethnic country in Chinese civilization III-1 Regime changes, ideological clashes, revolutionary processes in modern times III-2 Construction of the capital after the founding of the People's Republic of China IV-3 Local cultural traditions IV-4 Intellectual, cultural and technological exchanges between East and West	1. Protecting the Lugou Bridge (Marco Polo Bridge) and the Wanping Fortress and their historic settings; 2. Protecting important sites of the New Democratic Revo-lution, such as the site of the February 7th Strike in Changxindian; 3. Protecting modern industrial heritage sites, such as the Erq-i Locomotive Factory in Beijing; 4. Protecting significant sites exhibiting the traditional lives of old town residents, including the former Compound of the Feng Family, the Julaiyong Grocery Shop, and former First Barbershop in Changxindian
Shijingshan District	Moshikou-Baidachu-Shougang	II-1 Geographical pattern, providing the geographical foundation for the capital of a unified multi-ethnic country in Chinese civilization III-1 Regime changes, ideological clashes, revolutionary processes in modern times III-2 Construction of the capital after the founding of the People's Republic of China III-3 Great rejuvenation after the Reform and Opening-up IV-2 Diversity of religious beliefs IV-3 Local cultural traditions	1. Protecting religious heritage sites such as Cishan Temple and Xianying Temple; 2. Protecting important sites that exhibit traditional conne-ctions between urban and rural residents, such as the Moshi-kou Historic Conservation Area; 3. Protecting the important national memorial site of Babaoshan Revolutionary Cemetery; 4. Protecting significant sites related to modern industrial development such as Shougang, and the Winter Olympics legacies

Continued

District	Key Area	Core Value	Conservation Priority
Mengtougou District	Yanhecheng Fortress-Cuandixia Village-Donghulin Site	I-2 Origins of agriculture I-3 Geological features and biodiversity II-1 Geographical pattern, providing the geographical foundation for the capital of a unified multi-ethnic country in Chinese civilization II-6 The Great Wall defense system III-1 Regime changes, ideological clashes, revolutionary processes in modern times IV-3 Local cultural traditions	1. Protecting Neolithic sites such as the Donghulin Site; 2. Protecting the Yongding River and its surrounding wetland parks and forest parks; 3. Protecting ancient roads in western Beijing, such as ancient pilgrimage roads, ancient military roads, and ancient trade roads, as well as their associated transportation heritage properties; 4. Protecting important physical carriers that bear the Great Wall defense system, such as the Yanhecheng Fortress, as well as their surrounding environments; 5. Protecting revolutionary sites such as the former Headquarters of the Hebei-Jehol-Chahar Advance Army under the Eighth Route Army; 6. Protecting traditional villages, such as Cuandixia Village
Fangshan District	Jin Tombs-Peking Man Site at Zhoukoudian-Yunju Temple-Liulihe Ruins	I-1 Human origins I-3 Geological features and biodiversity II-1 Geographical pattern, providing the geographical foundation for the capital of a unified multi-ethnic country in Chinese civilization II-2 The history of Beijing as a city and a capital, bearing extraordinary testimonies to the development of a unified multi-ethnic country in Chinese civilization	1. Protecting important physical carriers of the Paleolithic Neolithic periods, such as the Peking Man Site at Zhoukoudian and the Zhenjiangying Site, and their surrounding environments; 2. Protecting the Protected Areas such as the Shangfangshan National Forest Park; 3. Protecting significant transportation heritage properties, such as the Liulihe Bridge; 4. Protecting important ancient city sites, including the Liulihe Ruins and the Doudian Site; 5. Protecting imperial mausoleums, noble burials, and their surrounding environments, including the Jin Tombs, the Prince Zhuang's Mausoleum, and the Manor and Mausoleum of Yihui and Gu Taiqing;

Continued

District	Key Area	Core Value	Conservation Priority
Fangshan District	Jin Tombs-Peking Man Site at Zhoukoudian-Yunju Temple-Liulihe Ruins	II-5 The most representative examples of ancient Chinese imperial architecture IV-2 Diversity of religious beliefs IV-3 Local cultural traditions	6. Protecting religious sites or cultural heritage properties in Fangshan, such as the Pagodas and Stone Sutras at Yunju Temple of Fangshan, the Wanfotang Hall, the Kongshui Cave Stone Carvings and Pagodas, and the Shizi Temple Ruins; 7. Protecting traditional villages such as Nanjiao Village and Shuiyu Village
Tongzhou District	Beijing Municipal Administrative Centre	II-1 Geographical pattern, providing the geographical foundation for the capital of a unified multi-ethnic country in Chinese civilization II-2 The history of Beijing as a city and a capital, bearing extraordinary testimonies to the development of a unified multi-ethnic country in Chinese civilization II-7 The canal system III-1 Regime changes, ideological clashes, revolutionary processes in modern times III-3 Great rejuvenation after the Reform and Opening-up IV-2 Diversity of religious beliefs IV-3 Local cultural traditions IV-4 Intellectual, cultural and technological exchanges between East and West	1. Protecting important physical carriers that exhibit the transformation of ancient regional administrative establishments and their surrounding environments, such as the Ancient Government Seat of Luxian County, Tongzhou Old Town, and Huoxian Old Town; 2. Protecting canal heritage properties, including the Grand Canal and related waterworks facilities along its route; 3. Protecting revolutionary sites such as the former headquarters of the Peiping-Tianjin Campaign; 4. Protecting major facilities planned or constructed in the Beijing Municipal Administrative Centre, including the Beijing Performing Arts Center, the Beijing City Library, the Beijing Grand Canal Museum, and the Central Green Forest Park; 5. Protecting significant sites related to the dissemination and development of religions, such as the Tongzhou Lighthouse Pagoda and the Ziqing Palace; 6. Protecting key areas such as the Eighteen Half-Cut *Hutongs*; 7. Protecting modern and cotemporary building sites, such as contemporary educational institutions complex in Tongzhou

Continued

District	Key Area	Core Value	Conservation Priority
Changping District	Juyongguan-Ming Tombs - Yinshan Pagodas-Baifuquan Site	II-5 The most representative examples of ancient Chinese imperial architecture II-6 The Great Wall defense system II-7 The canal system III-1 Regime changes, ideological clashes, revolutionary processes in modern times III-2 Construction of the capital after the founding of the People's Republic of China IV-2 Diversity of religious beliefs IV-3 Local cultural traditions	1. Protecting the Ming Tombs and its surrounding environment; 2. Protecting important physical carriers of the Great Wall defense system, such as the Juyongguan section of the Great Wall, and their surrounding environments; 3. Protecting modern railway heritage properties, such as the Nankou section of the Beijing-Zhangjiakou Railway; 4. Protecting canal heritage properties, including the Grand Canal (Baifuquan Site); 5. Protecting major projects in the construction of the capital, such as the Ming Tombs Reservoir; 6. Protecting religious sites or cultural heritage properties such as the Yinshan Pagodas; 7. Protecting traditional villages, such as Deling Village, Kangling Village, Maoling Village, and Wanniangfen Village
Daxing District	Nanyuan-Tuanhe Temporary Palace Site	I-3 Geological features and biodiversity II-5 The most representative examples of ancient Chinese imperial architecture III-1 Regime changes, ideological clashes, revolutionary processes in modern times III-2 Construction of the capital after the founding of the People's Republic of China IV-4 Intellectual, cultural and technological exchanges between East and West	1. Protecting important physical carriers of imperial gardens, such as the Tuanhe Temporary Palace Site, and their surrounding environments; 2. Protecting revolutionary sites such as the former site of the Nanyuan Barracks Headquarters (located in Fengtai District); 3. Protecting modern architectural heritage, such as the Nanyuan Airport

Continued

District	Key Area	Core Value	Conservation Priority
Huairou District	Huanghuacheng-Mutianyuan-Huairou Science City-Yanqi Lake	II-6 The Great Wall defense system III-3 Great rejuvenation after the Reform and Opening-up IV-1 Cultural interaction and integration among various ethnic groups IV-2 Diversity of religious beliefs	1. Protecting important physical carriers of the Great Wall defense system, such as the Mutianyu section of the Great Wall, and their surrounding environments; 2. Protecting characteristic areas that serve as hubs for modern international exchanges and the development of high-tech industries, such as the Yanqi Lake International Convention Center and the Huairou Science City; 3. Protecting important villages symbolizing multi-ethnic exchanges, such as Labagoumen Manchu Township and Changshaoying Manchu Township; 4. Protecting religious sites or cultural heritage properties, such as Hongluo Temple
Pinggu District	Jiangjunguan-Shangzhai Ruins	I-1 Human origins I-2 Origins of agriculture I-3 Geological features and biodiversity II-6 The Great Wall defense system III-1 Regime changes, ideological clashes, revolutionary processes in modern times	1. Protecting Paleolithic and Neolithic sites, such as the Shangzhai Ruins and the Majiafen Site; 2. Protecting the Protected Areas, such as the Jinhai Lake Scenic Area and the Huangsongyu National Geopark; 3. Protecting important physical carriers of the Great Wall defense system, such as the Jiangjunguan section of the Great Wall, and their surrounding environments; 4. Protecting revolutionary sites, such as the Yuzishan Anti-Japanese War Site

Continued Tabl

District	Key Area	Core Value	Conservation Priority
Miyun District	Gubeikou-Simatai-Baimaguan-Miyun Reservoir	I-3 Geological features and biodiversity II-6 The Great Wall defense system III-1 Regime changes, ideological clashes, revolutionary processes in modern times III-2 Construction of the capital after the founding of the People's Republic of China IV-3 Local cultural traditions	1. Protecting the Protected Areas, such as the Yunmengshan National Geopark and Gubeikou Forest Park; 2. Protecting important physical carriers of the Great Wall defense system, such as the Simatai section of the Great Wall, and their surrounding environments; 3. Protecting revolutionary memorial sites, such as the Gubeikou Battle Martyrs' Cemetery and the Bai Yihua Martyr Cemetery; 4. Protecting major capital construction projects, such as the Miyun Reservoir; 5. Protecting traditional villages, such as Huangyukou Village and Baimaguan Village
Yanqing District	Badaling-Guanting Reservoir-Venues of the Beijing 2019 International Horticultural Exhibition -Venues of the Beijing 2022 Winter Olympics	I-3 Geological features and biodiversity II-1 Geographical pattern, providing the geographical foundation for the capital of a unified multi-ethnic country in Chinese civilization II-6 The Great Wall defense system III-1 Regime changes, ideological clashes, revolutionary processes in modern times III-3 Great rejuvenation after the Reform and Opening-up IV-1 Cultural interaction and integration among various ethnic groups IV-4 Intellectual, cultural and technological exchanges between East and West	1. Protecting the Protected Areas, such as the Yeyahu Wetland Nature Reserve; 2. Protecting important physical carriers of the Great Wall defense system, such as the Badaling section of the Great Wall and the Chadaocheng Fortress Site, as well as their surrounding environment; 3. Protecting modern railway heritage properties, such as the Badaling section of the Beijing-Zhangjiakou Railway; 4. Protecting sites exhibiting lives of various ethnic groups, such as Yanqing Ancient Cliff Dwellings; 5. Protecting major facilities that support the "Four Centers" functions, such as the venues of the Beijing 2019 International Horticultural Exhibition and the Beijing 2022 Winter Olympics

(2) Overall strategy for the control of key areas

Enhance the interpretation and presentation of historical and cultural values, adhere to the principle of "balancing conservation and development", and strengthen overall control of the 14 key areas, ensuring their conservation while promoting sustainable development.

Protecting carriers of historical and cultural values. Uncover the core values carried by historic and cultural heritage. Ensure comprehensive conservation of fabrics of significant historic and cultural heritage properties and their surrounding environments, strengthening control over their historic features. Protect the physical carriers that embody these core values.

Enhancing the interpretation and presentation of the core values of each key area. Conduct in-depth research on the key areas, focusing on the historical origins, development trajectory, and main directions to explain their core values. For the physical carriers that reflect the core values, interpret their connotations and values with direct displays or supplementary means. For places that gain value through witnessing important historical events or famous individuals, respond to their core values in their functional and spatial planning, and set up markers for display and explanation.

Developing cultural theme routes linking the key areas. Explore the interconnections between the key areas, relying on linear spatial resources such as ancient paths, greenways, railways, rivers, ravines, and ridgelines, and create distinctive cultural theme routes based on their core values. Strengthen the coordinated conservation of historic and cultural heritage along these routes and select representative sections for focused display.

Improve the livelihood environment of the key areas. Increase the efforts to comprehensively rehabilitate the environment in the key areas, improve infrastructure, and effectively enhance the living conditions. Rely on the key areas to organize diverse cultural activities, enrich residents' lives and improve the quality of their lives.

Promote the economic development of the key areas. Leverage the key areas and cultural theme routes to strengthen the construction of infrastructure and service facilities related to heritage conservation and presentation and tourism development. Foster cultural exchanges and economic growth in the villages along the routes.

Chapter 4

Establishing an Inheritance Mechanism to Promote Chinese Culture

I   Deepening Value Exploration and Interpretation

1. Contributing to the development of the identity system of Chinese civilization

The origins of Chinese civilization and the historical process through which Beijing integrated into the diverse yet unified framework of Chinese civilization will be explored in a thorough and multifaceted manner. The essence of ancient capital culture and the unique cultural flavor of Beijing will be uncovered. Interdisciplinary and cross-sectoral innovative research will be enhanced. This includes studying the significance of the Taihang and Yanshan Mountains ranges in the formation of Beijing as a capital and the unification of the country, providing a holistic interpretation. Through archaeological evidence, key issues such as the origins of human activities, the origins of agriculture, urban development, and the exchanges and mutual learning between diverse civilizations in Beijing will be explored. Ongoing excavation activities will be carried out at major archaeological sites, such as the Peking Man Site at Zhoukoudian, Donghulin Site, Liulihe Ruins, the Ancient Government Seat of Luxian County, Site of the Jin Tombs, Site of Zhongdu of the Jin Dynasty, as well as at the Old City of Beijing, the Yuanmingyuan, and the Great Wall. Efforts will be made to conduct regional archaeological surveys along the Yongding River and its tributaries, building a framework for the evolution of settlement patterns and clarifying the prehistoric cultural evolution in Beijing. Archaeological mapping, surveys, exploration, and excavation will continue, alongside the establishment of a platform for basic archaeological information. Digitization and vectorization of archaeological research findings will be advanced.

Research on Beijing's historic and cultural heritage that carry core values and possess significant meaning and important influence will be strengthened. The historical stories, cultural values, and spiritual connotations of these heritage sites will be systematically excavated, sorted, and presented. Identity symbols that can represent the national image and the unique spirit of the Chinese nation will be selected to develop the Beijing chapter of the identity system of Chinese civilization. The conservation and inheritance of eight World Cultural Heritage Sites, including the Great Wall, the Forbidden City in Beijing, the Peking Man Site at Zhoukoudian, the Summer Palace, the Temple of Heaven, the Ming Tombs, the Grand Canal, and Beijing Central Axis, will be strengthened. Eligible properties will be recommended for inclusion in the Tentative List for World Heritage nomination.

2. Intensively elaborating the great journey of socialist modernization in the capital city

The heritage listings of various types of protected properties, such as revolutionary sites, historic buildings, and areas of special quality, will be continuously enriched and expanded. The inheritance and promotion of revolutionary culture and innovative culture will be vigorously carried out. Based on the study of histories of the Communist Party of China, the People's Republic of China, the Reform and Opening-up, and socialist development, the important construction achievements scored by the Party and the people during various historical periods will be preserved. Representative heritage properties exhibiting the spirit of the May Fourth Movement, the spirit of "Entering Beijing to face New Tests", the spirit of "Two Bombs, One Satellite", the model worker spirit, the labor spirit, the craftsmanship spirit, the scientist spirit, and the Beijing Winter Olympics spirit, which are key elements of the spiritual lineage of Chinese Communists, will be systematically summarized and their contemporary value highlighted.

II   Intensifying Efforts of Cultural Heritance in Urban and Rural Construction Practices

1. Properly protecting and inheriting the spatial framework of the ancient capital

The conservation and inheritance of historical and cultural heritage in urban and rural areas will be guided by the Master Plan, respecting and transmitting the urban genes of excellent traditional planning and design, and promoting the integration of the conservation of historic and cultural heritage in urban and rural areas with planning and construction practices.

Systematic research on traditional concepts and practices for the planning and construction of the ancient capital of Beijing will be strengthened, inheriting and promoting the magnificent spatial framework of mountains and rivers surrounding the city, orderly structure, and the fusion of rites and music. The importance of the Old City and the Three Hills and Five Gardens as the capital's central functions is emphasized, maintaining the historical layout. The leading role of the "Two Axes" in the city's spatial framework will be continued, reserving key cultural and international exchange functions for the nation at important nodes along the axes. The large-scale mountain-and-river-based pattern will be highlighted and an integrated "mountain-river-city" order created, emphasizing the significant role of the natural environment in urban and rural construction.

2. Foster urban spaces with cultural characteristics

The guidance on urban design will be strengthened to highlight regional features and landscapes. An urban spatial order framework will be developed, strengthening regulation over architectural features, coordinating and harmonizing spatial elements such as building height, volume, color and the fifth façade.

The development of public environmental art will be promoted, building a cultural environment that aligns with historical traditions. The historical information and cultural connotations behind public spaces should be fully explored, improving the level of public space design by incorporating cultural elements and fully showcasing their humanistic significance. The construction of an urban cultural gene system should be strengthened, and the city's visual identity system should be reinforced to highlight regional characteristics.

The continuity of contemporary architectural culture should be promoted, creating modern buildings rooted in historical environments. Wisdom from traditional architecture should be drawn upon to revitalize traditional building culture, advancing its modern inheritance and application. A blend of Chinese and foreign influences should be embraced to elevate contemporary architectural design theories and methods, creating national architectural masterpieces worthy of history and the times, showcasing an elegant style that harmoniously coexists with historical spatial order. The review of architectural designs in key cultural heritage areas should be strengthened.

Control over the natural environment and village landscapes in rural areas should be enhanced. The use of locally distinctive building materials and construction methods should be encouraged for village development, protecting and inheriting cultural elements, architectural characteristics, and construction features with cultural lineage, better reflecting regional characteristics. The connection between villages and natural landscapes should be strengthened, creating village spaces with cultural depth.

3. Strengthening the conservation and inheritance of cultural heritage in urban renewal and renovation

The conservation of historic and cultural heritage in urban renewal should be effectively strengthened, with the principle of preservation being prioritized while a combination of preservation, renovation and demolition is adopted. Before the renewal and renovation of old neighborhoods, residential areas, towns, villages and factory areas, prior assessment of historical and cultural value should be conducted, protected heritage properties should be identified and publicly announced in a timely manner, and conservation measures should be implemented.

The *Regulations of Beijing Municipality on Urban Renewal* should be implemented to inherit and continue the traditional urban fabric, preserving the unique regional environment, cultural characteristics and architectural styles of the renewal areas. Actions that damage the terrain and topography, harm or cut down ancient and notable trees, or alter the historical layout and environmental appearance are prohibited. Large-scale demolition and construction, the practice of demolishing authentic structures and rebuilding fake ones, and the confusion of authenticity with inauthenticity should be strictly prohibited. The indiscriminate demolition of old buildings, old residences, old archways, ancient wells and old bridges with conservation value should not be carried out. Existing buildings, spaces and landmarks that reflect specific stages of urban development, important historical events and public emotional memories should be effectively protected. New and renovated projects should continue the traditional urban fabric and showcase the city's unique characteristics.

Ⅲ Strengthening the Conservation and Inheritance of Cultural Heritage

1. Establishing a value-centered conservation and utilization mechanism

The conservation and utilization of cultural heritage should be centered around highlighting its historical and cultural value, with a focus on value interpretation and the inheritance of cultural spirit. Different protected heritage properties should be approached with more open and flexible conservation and utilization models, aiming to address the imbalance and inadequacy in the conservation and utilization process.

Conservation and utilization must be premised on ensuring the safety of the heritage itself. No unauthorized destruction or damage to protected heritage properties, or actions that impact the environmental appearance, should be allowed. The approach should align with the heritage's intrinsic value, prioritizing the service of public welfare functions.

2. Increasing the accessibility of built heritage

It should be emphasized that conservation be promoted through utilization, with efforts fully made to improve the conservation and utilization of built heritage, such as historic monuments, historic buildings and historic gardens, especially ancient buildings and representative buildings in modern times. The goal is not to aim for ownership, but for conservation, with open access to the public. Active efforts should be made to relocate and protect key historic monuments and historic buildings that are currently being used inappropriately, achieving either full or moderate access. It should be encouraged to open non-government-owned historic monuments, historic buildings and historic gardens to the public and provide exhibition services, with relevant authorities offering guidance and support.

The historical functions of historic monuments, historic buildings and historic gardens should be respected. In accordance with relevant regulations such as the *Guidelines of Beijing Municipality for the Accessibility and Utilization of Historic Monuments (Trial), Standards for the Conservation and Utilization of Historic Monuments – Former Residences of Famous People, Guidelines for the Presentation of Revolutionary Sites*, and *Interim Administrative Measures for the Conservation of Industrial Heritage in Beijing*, and taking into account their historical and cultural values, these sites should be used as venues for state affairs, offices for public purposes, or places for cultural exhibitions or tourism and leisure activities.

Exploration of policy innovations such as function replacement, compatible use, rental discounts and financial subsidies is encouraged, with function transformation being guided by demand to promote the unity of conservation and utilization. Due to the needs for conservation and inheritance, value interpretation, as well as the improvement of urban functions and addressing urban deficiencies, the use of historic monuments, historic buildings and historic gardens may be converted after the corresponding approval procedures are followed in accordance with laws and regulations. The converted use must comply with the conservation requirements and value features of protected heritage properties.

The promotion of cluster-based conservation and utilization is encouraged by leveraging "cultural IP" and "city brands" to diversify cultural business formats and extend the value chain.

3. Improving the living environment of block-based cultural heritage

The infrastructure and public service of historic conservation areas, historic towns and villages, and traditional villages should be improved in accordance with local conditions. Disaster prevention and mitigation, as well as sanitation facilities, should be strengthened to enhance the quality of the living environment. An implementation mechanism for heritage conservation, with residents as primary practitioners, should be established. Efforts should be made to promote application for termination of tenancy or rent exchange, protective repair, and restorative construction, demolishing illegal structures within courtyards and restoring the basic layout of traditional *siheyuan* compounds. On the basis of effective conservation, the residential environment should be improved, vacated houses should be utilized rationally and efficiently, and the renovation of courtyard houses should be promoted. Elder-friendly facilities and kitchen and bathroom facilities should be added, allowing for the integration of cultural heritage and modern life, and restoring the living scenes with an old Beijing flavor. Efforts should be made to explore and establish a property service model suitable for the actual situation of courtyard areas and promote the autonomous renovation of courtyard houses.

Multiple measures should be taken to explore new approaches for the functional renewal of historic conservation areas, historic towns and villages, and traditional villages. The stock update mechanism for courtyard areas should be innovated, promoting the coordination of darning historic features and leveraging assets, strengthening synergy between co-building communities and weaving historic fabrics, and enhancing the conservation and regeneration of historic conservation areas. Small but refined, distinctive historic streets, alleys and neighborhoods should be cultivated, enriching cultural and tourism consumption scenarios, expanding new urban consumption spaces, and driving regional revitalization. Traditional villages should be encouraged to implement more open industrial development policies, support the development and expansion of collective economies in historic towns and villages and traditional villages, and explore ways for rural collective economic organizations and villagers to participate in the conservation and development of traditional villages through equity contributions using the land use rights for collective construction land for business purpose, idle homestead land, houses, etc. The concentrated and contiguous conservation and development of traditional villages in Mentougou, Miyun, Fangshan, Changping, Yanqing, Huairou and Pinggu should be actively promoted, aiming to retain local residents, protect local heritage, and preserve local memories. Efforts should be made to strengthen the integration of traditional farming culture and traditional village conservation, and actively promote the nomination of sites where agricultural heritage systems are located for the protective designation of traditional villages.

4. Strengthening value interpretation of historic sites

Cultural connotations of ancient sites, underground archaeological remains and city-site remains should be explored. Value of important paleoanthropological sites, city-site remains, large architectural complexes and garden sites, mausoleums and large burial groups, and revolutionary sites should be comprehensively evaluated. Comprehensive efforts of conservation, inheritance and utilization should be carried out with focus on highlighting their national attributes and the development of the national political center, cultural center, and center for international exchanges, aiming to establish national and municipal memorial sites and build public cultural facilities such as site parks, museums and memorial halls.

Priority should be placed on the construction of archaeological site parks. Overall arrangement for archaeological site parks across the city should be planned. National archaeological parks of Zhoukoudian and Yuanmingyuan sites should be upgraded. The construction of the Liulihe National Archaeological Site Park and the Archaeological Site Park of the Ancient Government Seat of Luxian County should be prioritized, while the construction of archaeological site parks at Shangzhai, Donghulin, Zhongdu of the Jin Dynasty, and the Jin Tombs should be advanced.

Priority should be placed on urban archaeology. Archaeological activities on city-site remains from various historical periods of Ji, Tang, Liao, Jin and Yuan as well as their conservation and presentation should be carried out systematically. Based on archaeological projects at the Great Wall, the Grand Canal, the Central Axis, and Yuanmingyuan, thematic cultural interpretation of the relationship between historic sites and remains and urban public spaces. should be strengthened.

Priority should be placed on modern revolutionary sites. A number of key landmark locations around significant events and moments should be identified, with an aim to build spiritual shrines for Chinese communists, spiritual homeland for Chinese people, and spiritual high ground for the Chinese nation.

5. Revitalizing intangible cultural heritage and time-honored brands

Comprehensive safeguarding of intangible cultural heritage should be strengthened, integrating it into daily life and production, and ensuring space for its transmission. For the intangible cultural heritage that embodies exquisite craftsmanship, such as traditional techniques, traditional arts, and traditional medicine, transmission bases that combine craft display with product creation should be established, based on places for practicing intangible cultural heritage. For different forms of performing arts, such as traditional opera, traditional dance, traditional sports, games and acrobatics, immersive showcases should be presented in cultural spaces like historic monuments, historic buildings, and historic gardens. For the folk customs and intangible cultural heritage items with local characteristics that have been created during the development of each village, should be integrated into the growth of villages to contribute to rural revitalization. Integration of intangible cultural heritage with tourism should be promoted to enrich Beijing's cultural and tourism resources.

Preservation of time-honored brands at their original locations and in their original appearance should be strengthened. The *Action Plan for Further Promoting the Innovative Development of Beijing's Time-honored Brands (2023-2035)* should be actively implemented. Time-honored brands are encouraged to establish museums, galleries, experience centers, and cultural houses to diversify methods of their use. Featured products and services of time-honored brands should be upgraded to enhance their prestigious reputation and inspire their internal vitality.

6. Contributing to sustainable urban and rural development

The important role of the conservation and utilization of cultural heritage in value promotion, environmental improvement, social participation, and public education should be highlighted. Coordination between cultural heritage conservation and surrounding urban and rural development should be strengthened. Urban and rural spatial layout should be optimized to improve functions and enhance vitality. The shaping of characteristic cityscape and restoration of urban ecology should be reinforced to continue the historical context of the city and support sustainable urban and rural development. The role of large-scale historic sites, historic conservation areas, and linear heritage properties in optimizing regional spatial patterns should be enhanced. Culture should be leveraged as a link to connect heritage, ecology, industry, cities and villages to drive the transformation and upgrading of regional economy.

7. Supporting major national strategies

The construction of national culture parks should be promoted. With the construction of national culture parks as a key focus, the functional layout of urban and rural areas should be optimized, and the development of the surrounding regions should be promoted. A comprehensive conservation framework should be established, connecting heritage sites in a continuous and linear manner, shaping urban and rural landscapes that integrate historical context, ecological environment, and modern facilities.

The establishment of national "cultural landmarks" in Beijing should be promoted. Focusing on the cultural development needs of the new era, the construction of major modern public cultural facilities with national representation, strong radiating influence, significant meaning, and important impact, or those that can play a key role in international cultural exchanges, should be encouraged.

The implementation plan for patriotic education should be thoroughly carried out. Focus should be placed on key historical moments such as the founding of the Party, the military, and the nation, as well as the central tasks of the Party and the State. Relying on revolutionary heritage, a series of commemorative activities and mass theme-based education programmes should be organized.

IV Promoting the Innovative Presentation and Dissemination of History and Culture

1. Enhancing the presentation of history and culture

The important natural and cultural landscape resources should be leveraged, with value themes as the main focus, to deeply explore historical stories. Various historical and cultural resources should be utilized as the backbone, supplemented by scenic areas, natural landscapes, and cultural venues. Linear heritage properties such as canals, the Great Wall, railways, and ancient roads should be organically connected and strung together, creating cultural theme routes.

Major cultural theme routes with Chinese cultural identity should be developed to increase the international visibility of the capital's culture, in conjunction with the construction of the Great Wall and Grand Canal National Culture Parks, the conservation and inheritance of the Central Axis, the environmental improvement of Chang'an Avenue, the establishment of the Three Hills and Five Gardens National Cultural Heritage Conservation and Utilization Demonstration Area, and the preservation and inheritance of revolutionary culture. Districts should be encouraged to refine and shape district-level cultural theme routes that display regional cultural characteristics, based on historic roads, mountains and rivers, as well as historic sites, events and figures.

Cultural theme routes should be developed by following a value-oriented approach, with a focus on value interpretation. The philosophical ideas, humanistic spirit, and value concepts behind these routes should be revealed, emphasizing authenticity and public nature, while avoiding excessive commercialization and entertainment. Methods to use cultural theme routes should be diversified, encouraging innovative activities such as study tours, hiking, cycling, self-driving camping, folk performances, and sports events. The integration of cultural theme routes with the tourism industry should be actively promoted, and development of ecological tourism, boutique homestays, forest wellness, and leisure agriculture should be advanced.

2. Strengthening the cultural education system

The integration of historical and cultural education into schools should be continuously promoted. Preservation and inheritance of history and culture should be comprehensively incorporated into all stages of education, including early education, basic education, vocational education, higher education, and continuing education. The establishment of disciplines and professional programmes related to the preservation and inheritance of history and culture should be promoted in higher education institutions and vocational schools. Relying on the rich cultural heritage resources in Beijing's urban and rural areas, various activities such as educational tours of history and educational practices should be encouraged.

The cultivation of a talent pool for the conservation and inheritance of Beijing's cultural heritage should be promoted. An approach to combine talent cultivation and talent discovery should be followed, aiming to build a high-level platform for the conservation and inheritance of Beijing's cultural heritage. A sound training and evaluation system for intangible cultural heritage bearers and traditional craftsmen should be established. Support should be provided to grassroots key figures, and efforts should be made to nurture local cultural talents and folk culture inheritors rooted in communities.

3. Serving the construction of the national cultural center

The role of the conservation and inheritance of cultural heritage in leading spatial development for the construction of the national cultural center should be fully utilized. Efforts should be made to enhance the cultural narrative of the capital, promoting the continuous integration and innovation of fine traditional Chinese culture, revolutionary culture, and advanced socialist culture, achieving more fruitful cultural outcomes.

Focusing on the culture of the ancient capital, revolutionary culture, Beijing flavor culture, and innovation culture, the advantages of Beijing's rich cultural heritage and concentrated cultural resources should be leveraged. The branding and development of "City of Museums", "City of Books", and "Capital of Performing Arts" should be continuously advanced, creating cultural brands with distinct characteristics for the city.

Platforms for cultural interaction and presentation in various types and at various levels should be established, attracting excellent cultural works from across the country to be displayed and exchanged in the capital. Restored and repaired venues such as guild halls and former residences of famous figures should be utilized to provide spaces for showcasing cultural masterpieces from other areas.

4. Enriching the content and forms of historical and cultural supplies

Integration into production, daily life, and artistic creation should be promoted. Various cultural theme activities, such as traditional festivals and commemorative events, should be organized. Significant historical milestones, such as the founding of the city and the capital, should be linked with memorial days to deepen the public's understanding of fine culture.

Innovative methods of promotion and publicity should be employed. The role of newspapers, books, radio, the internet, and new media platforms should be fully utilized. News reports, TV dramas, television programs, documentaries, cartoons, short videos, and other formats should be encouraged to promote the involvement of individuals, social organizations, and enterprises in cultural promotion, thereby broadening publicity channels. Activities such as "History and Culture into Government Agencies", "History and Culture into Schools", and "History and Culture into Communities" should be continuously carried out to enhance the visibility, cohesion, and appeal of Beijing's history and culture.

5. Promoting the global spread of Beijing's history and culture

State event venues will be constructed. For the landmarks of Chinese civilization that can highlight fine traditional Chinese culture and meet corresponding conditions in terms of scale and location, functions for national ceremonial activities and state events will be planned in order to promote exchanges and mutual learning among civilizations.

International exchanges and cooperation will be deepened. Exchanges and cooperation with countries involved in the Belt and Road Initiative will be strengthened, and the sharing of experiences in the preservation and inheritance of history and culture will be promoted. Historical and cultural venues that meet the necessary conditions will be actively supported to host appropriate international exchange activities. Renowned cultural events such as the Beijing Culture Forum, Beijing International Design Week, China (Beijing) International Canal Arts Fest, and Beijing International Cultural & Creative Industry Expo will be enhanced, with the aim of creating a top-tier platform for cultural exchanges with global impact.

The capacity for international communication will be enhanced. Adhering to the positioning of the capital, national perspective, and global vision, overseas platforms such as China Cultural Centers and Confucius Institutes, along with domestic and international expos, will be fully utilized to tell the story of China and the story of Beijing. The "going global" cultural efforts will be continuously advanced, with the unique role of history and culture in supporting national diplomacy being highlighted, thereby enhancing the international visibility of Beijing's cultural heritage.

V Strengthening the Collaborative Conservation of Cultural Heritage in the Beijing-Tianjin-Hebei Region

1. Establishing a collaborative conservation mechanism

Top-level design will be strengthened to collaboratively push forward legislative work for formulating policies for the conservation of cultural heritage in the Beijing-Tianjin-Hebei region as well as the conservation of the Yongding River. Efforts will be made to establish a regular communication mechanism for the conservation of cultural heritage in the Beijing-Tianjin-Hebei region, enhancing communication and consultation on policies and measures related to cross-regional cultural heritage. The establishment of collaborative research platforms will be encouraged, focusing on important cultural heritage sites, such as the Great Wall, the Grand Canal, the Imperial Tombs of the Ming and Qing Dynasties, the Yan capital sites, and ancient human sites, in order to elevate the level of collaborative research on history and culture in the Beijing-Tianjin-Hebei region. Cooperation between universities in Beijing, Tianjin and Hebei will be promoted, and joint research and training in the conservation and inheritance of cultural heritage will be carried out.

2. Improving the conservation and inheritance system for cultural heritage in the Beijing-Tianjin-Hebei region

The conservation and inheritance of cultural heritage will be prioritized as a key task and strategic support for promoting the coordinated development of the Beijing-Tianjin-Hebei region. The cultural characteristics of the region will be refined, fully showcasing the significant role that the Yan-Zhao region and the capital area have played in the history of Chinese civilization, the history of the Communist Party of China, the history of the People's Republic of China, the history of reform and opening-up, and the history of socialist development. A holistic conservation approach will be followed to systematically advance the conservation of historic cities, towns and villages, traditional villages, and historic conservation areas in the region, thereby maintaining the historical and cultural continuity of urban and rural areas. World Heritage sites in the region will be comprehensively protected, developing landmarks of Chinese civilization and encouraging joint nomination efforts for World Heritage status. The conservation of large-scale archaeological sites and the construction of national archaeological site parks will be advanced. The comprehensive conservation of the mountain-and-river-based pattern of the Yanshan and Taihang mountains will be ensured, promoting the integration of the preservation of history and culture, the conservation of natural landscape, and the urban and rural development. The implementation of intangible cultural heritage transmission and development programmes will be deepened, exploring possibilities to establish cultural ecosystem conservation areas, improving the systematic safeguarding and transmission of intangible cultural heritage. Various types of industrial and agricultural heritage, including railways, mining and metallurgical sites, and ports, will be protected to retain the historical memory of industrial and agricultural production and construction activities. Maritime cultural resources and heritage will be further explored to celebrate the nation's maritime culture.

The conservation and development of the Great Wall Cultural Belt, the Grand Canal Cultural Belt, the Taihang Mountain Cultural Belt, and the Bohai Bay Maritime Cultural Belt will be coordinated. National strategies will be implemented, with collaborative efforts made in the construction of the Great Wall and Grand Canal National Culture Parks. Feasibility studies will be jointly conducted for the nomination of the Taihang Mountain National Culture Park. The historical, cultural and ecological resources along the Great Wall, the Grand Canal, Taihang Mountain, and Bohai Bay will be protected, preserved and utilized. Key projects, including cultural heritage conservation and inheritance, ecological protection and restoration, improvements in people's livelihoods, and the integration of culture and tourism, will be systematically advanced to enhance the effectiveness of cultural heritage conservation, inheritance and utilization, establishing cultural landmarks that are both national and global in significance. The revolutionary culture will be inherited and promoted, fully exploring resources of the revolutionary culture related to the May Fourth Movement, the War of Resistance against Japanese Aggression, the Pingjin Campaign, the "Entering Beijing to face New Tests" event, and the founding of the People's Republic of China, forming classic sightseeing routes to experience the revolutionary culture in the Beijing-Tianjin-Hebei region.

Chapter 5

Improving the Implementation Mechanism to Ensure Effective Management

I　Improving the Work Mechanism

1. Strengthening planning and coordination

The leadership of the Party shall be effectively strengthened. A work mechanism for the conservation and inheritance of cultural heritage will be established and improved, with Party committees providing leadership, the government playing a coordinating role, entities implementing tasks, and public participation and supervision ensured. Party committees and governments at all levels will fully implement the decisions and arrangements of the CPC Central Committee and the State Council, and recognize the significant importance of strengthening the conservation and inheritance of cultural heritage in urban and rural development. The functions of the Beijing Historic City Conservation Commission (hereinafter referred to as the "Commission") in overall planning, coordination, promotion, and supervision will be strengthened, and the work mechanism of the Commission and its Office for advancing the conservation and inheritance of cultural heritage will be improved. Relevant requirements for reporting and approval of major matters will be enforced. The conservation and inheritance of cultural heritage will be incorporated into training courses for officials, enhancing the awareness and capacity of Party and government leaders at all levels regarding the preservation and inheritance of history and culture in urban and rural development, and firmly establishing the concept that heritage conservation is also a measure of governance achievement.

Vertical coordination and cross-sectoral synergy should be strengthened. The city-wide unified approach should be reinforced, ensuring effective central-local collaboration, inter-regional coordination, and city-district synergy, while building an integrated management and service system. Efforts should be made to strengthen planning and coordination, policy and institutional alignment, and resources sharing. The role of local governments should be fully leveraged to improve district-level conservation and inheritance mechanisms, and to enhance the overall coordination, organization and implementation of conservation and inheritance work in key areas such as the Old City and the Three Hills and Five Gardens. The coordinated conservation of cultural heritage and the ecological environment should be strengthened, exploring a multi-departmental coordination and joint law enforcement mechanism.

The important role of grassroots governments in conservation and inheritance should be effectively highlighted. Efforts should be made to shift the focus of conservation management downward, establishing an implementation mechanism for the preservation of history and culture characterized by district-level coordination, townships performing main duties, inter-departmental collaboration, professional support, and broad public participation. The strength of designated area planners should be fully utilized to provide foundational support for district-level conservation efforts. Continuous strengthening of the workforce for cultural heritage conservation at all levels should be pursued, enhancing the capacity and proficiency of grassroots conservation teams.

2. Strengthening the transmission of planning

The vertical transmission of planning should be strengthened. Requirements of the State will be implemented, and upon approval, the plans will be integrated into the Integrated Map for Natural Resources Management and Territorial Spatial Planning. The transmission and coordination of major specialized conservation plans, including those for historic cities, towns and villages, traditional villages, and historic conservation areas, as well as the conservation and development plans for the three cultural belts, will be ensured to guide preservation and inheritance efforts in each district.

Horizontal coordination in planning should be strengthened. Efforts will be made to deepen the integration of the Master Plan for Conservation and Inheritance of Beijing's Historic and Cultural Heritage with other relevant plans, such as the national economic and social development plan. For areas with rich and concentrated historical and cultural resources, as well as areas where intangible cultural heritage is highly dependent on the natural environment and historical and cultural spaces, clear spatial control requirements for comprehensive conservation and utilization will be set out.

3. Promoting multi-party participation

Exploring various approaches for public participation in cultural heritage conservation. A broad social participation mechanism will be established, and channels for participation will be opened up to provide platforms for businesses, higher education institutions, research institutes, social organizations, volunteers, and interested individuals to engage in the conservation and inheritance of cultural heritage. Diverse stakeholders will be encouraged to play an active role in the planning, development and management stages of cultural heritage conservation and inheritance. The system of responsibility for conservation will be reinforced, defining the duties and obligations of owners and users of protected elements. The business environment will be further optimized, approval processes streamlined, preferential policies developed to encourage market entities to continuously engage in the conservation and inheritance of cultural heritage.

Improving the expert consultation mechanism. The development of the Expert Think Tank will be continuously advanced, with specialized research conducted on key issues related to cultural heritage conservation and inheritance. The consultation and evaluation mechanisms will be improved, enabling experts to play a vital role in decision-making, business consultation, and work guidance. An expert database at the district level, covering various fields such as land and space planning, historic city conservation, cultural heritage preservation, architectural engineering, historical and human geography, and archaeology, will be established. The mechanism for expert guidance at the grassroots level will also be enhanced.

Optimizing the reward and incentive mechanism. Research will be conducted to develop reward and subsidy policies, supporting the conservation and inheritance of cultural heritage in urban and rural areas through methods such as awards in lieu of subsidies and financial grants. The collection of demonstration cases for heritage conservation and inheritance will be continuously carried out, with timely summaries of successful experiences and practices. Organizations and individuals who make outstanding contributions to heritage conservation and inheritance will be recognized and rewarded in accordance with relevant national and local regulations.

4. Strengthening supervision and inspection

The requirements set out in the *Guidelines on Strengthening the Preservation and Inheritance of Historical and Cultural Heritage in Urban and Rural Construction* will be implemented, and the supervision and inspection mechanisms will be improved. Strict law enforcement will be enforced, with enhanced inspections and the establishment of a comprehensive evaluation mechanism, encouraging social oversight.

Law enforcement efforts will be intensified to strictly address violations. Under the leadership of district governments, sub-district offices and township governments will strengthen patrols of the state of conservation of cultural heritage within their jurisdictions, promptly addressing actions that endanger protected elements and reporting to relevant authorities. The oversight of cultural heritage conservation will be reinforced, ensuring the effective implementation of the "dual responsibility" system for both the party and government, as well as "dual duties" for individual roles, with strengthened supervision, inspections, and accountability for results.

Improving the regular evaluation and assessment mechanism. Efforts will be made to strengthen the survey and evaluation of historical and cultural resources, expanding the list of protected elements in a timely manner. The "Annual Check-up and Five-Year Evaluation" mechanism will be implemented, aligning with the overall urban planning evaluation and other relevant assessments. Specialized evaluations will be conducted for the city as a historic city, reviewing the implementation of protection and management requirements by relevant departments, as well as the conservation and inheritance of various protected elements. The evaluation results will serve as an important basis for improving government service efficiency and adjusting policies. The conservation and inheritance of cultural heritage will be incorporated into the evaluation system for civilized urban areas. Public interest litigation related to the conservation and inheritance of cultural heritage in urban and rural areas will be strengthened.

Encouraging public participation and supervision. Cultural heritage administrative information will be made more transparent, with public access to cultural heritage conservation and inheritance information services, allowing for greater social oversight. Citizens, legal persons, and other organizations will be encouraged to report any violations of laws or regulations concerning cultural heritage conservation and inheritance.

Ⅱ Improving the Guarantee Mechanism

1. Improving the supporting policy system

A comprehensive, full-process management approach will be followed, in line with the latest requirements of the State for cultural heritage conservation and inheritance. The legal framework will be strengthened, and the supporting policy system for protection and management will be improved, ensuring the management of the entire conservation and utilization process. The approval and administration procedures for the protection, repair, renovation, relocation and utilization of various protected elements will be optimized, with enhanced supervision during and after the process.

The timely modification of local regulations for the conservation of cultural heritage will be promoted. Multiple approaches will be explored for the vacation, protection, utilization and public display of historic monuments. Policies for the protection and management of traditional villages will be advanced. Supporting policy mechanisms for application for rent exchange (or relocation), protective repair, and restorative construction will be further improved. A baseline management model for protection and utilization will be explored, with project entry positive and negative lists established by type and area, periodically assessed, and dynamically adjusted. Policy coordination will be strengthened, and more effective and practical support policies for utilization, covering areas such as planning, fire safety, environmental protection, and certification, will be developed. A comprehensive building management system will be established, integrating the reform of the engineering project approval system, with stronger regulation of the renovation and demolition of existing buildings.

2. Strengthening financial support

A robust fiscal support mechanism for cultural heritage conservation and inheritance will be established, ensuring the effective use of municipal conservation funds. Priority will be given to supporting national and municipal-level projects that highlight the core values of Chinese civilization and Beijing's cultural heritage. The management of conservation fund transfers will be standardized, enhancing the autonomy and initiative of districts in their conservation efforts and ensuring the proper matching of rights and responsibilities. The management of cost performance will be strengthened, ensuring the effective utilization of fiscal funds. National support policies will be fully utilized, with a focus on leveraging the guiding role of central government specialized fiscal grants. The development of funding and subsidy mechanisms for the repair of historic towns and villages, traditional villages, and historic buildings will be explored.

Social capital will be encouraged to participate in heritage conservation efforts. Financial support will be increased based on market principles, and market-oriented methods and operations will be fully leveraged to effectively diversify funding channels. A new model for heritage conservation will be gradually shaped, one in which the government, society, residents, and other stakeholders collaborate, each contributing resources in a cooperative manner.

3. Advancing digital empowerment

Efforts will be accelerated to build cultural heritage conservation and management information platforms, and digital technical standards for various protected elements will be improved. Digital information collection and archiving will be conducted, enhancing interactive digital displays, with the gradual realization of digital management. The integration and sharing of data across multiple departments will be strengthened, establishing a monitoring indicator system for cultural heritage conservation. Multi-source data will be dynamically collected and integrated, utilizing the Integrated Map for Natural Resources Management and Territorial Spatial Planning to enable real-time control, dynamic monitoring, regular assessment, and timely early warning for heritage conservation and utilization. Satellite remote sensing technology will be employed to monitor construction activities within the protected areas of cultural heritage. Platforms for public participation and open access to information on cultural heritage will be developed to advance the modernization of spatial governance system and governance capacity. Modern technological methods will be used for the digital protection, processing and display of cultural heritage, with pilot projects for smart technology applications and innovative approaches for disseminating historical and cultural content.

4. Establishing a safety and emergency management mechanism

Strengthening the security guarantee mechanism. A prevention-first approach will be upheld, with an emphasis on daily maintenance and the establishment of a sound monitoring and early-warning system. Safety risk assessments and inspections for cultural heritage will be intensified, with improvements in safety engineering and risk resistance. Immediate reinforcement and repair will be carried out for heritage properties at risk of damage to eliminate safety hazards. A comprehensive disaster prevention system will be improved, with enhancements in fire safety, fire-fighting capabilities, and the implementation of anti-seismic and disaster-reduction engineering projects. A comprehensive prevention approach, utilizing human, material and technological resources, will be adopted to improve the disaster prevention, mitigation, and rescue capabilities for various cultural heritage sites.

Establishing a sound emergency management mechanism. A joint command and response mechanism will be set up for emergencies related to heritage safety and new archaeological discoveries, focusing on information sharing during emergencies, urgent sensitive event consultations and evaluations, as well as strengthening expert guidance and support to enhance the safety guarantee for cultural heritage.

Ⅲ Strengthening Project Management

The construction of a project database for the conservation and inheritance of cultural heritage will be further enhanced, with a focus on highlighting the core value of cultural heritage. Key projects will be considered and reserved in advance, and projects that are ready for implementation will be incorporated into budget arrangements according to the maturity.

The mechanism for pilot of projects will be actively promoted. Pilot projects will be carried out to address policy bottlenecks, implementation difficulties, and other critical issues. A communication and coordination mechanism for pilot projects will be established, and support, guidance and supervision for these projects will be strengthened. Dissemination and application of best practices from pilot projects shall be encouraged with favorable policies, formulation of standards and supporting funds.

附图

Annexed Maps